TECHNIQUES FOR TECHNOLOGY TRAINING

SHARON E. JOHNSON-ARNOLD

EDITED BY MARY STEARNS SGARIOTO

authorHOUSE®

AuthorHouse™
1663 Liberty Drive
Bloomington, IN 47403
www.authorhouse.com
Phone: 1-800-839-8640

First published by AuthorHouse 6/10/2010

ISBN: 978-1-4490-1810-8 (e)
ISBN: 978-1-4490-1809-2 (sc)

Library of Congress Control Number: 2010905846

Printed in the United States of America
Bloomington, Indiana

This book is printed on acid-free paper.

CONTENTS

Chapter 8. Launching the Training Day 99

Chapter 9. Advice from Training Professionals 115

PREFACE

I love watching any movie or TV show related to education. A few come to my mind, but my absolute favorite is Barbra Streisand in the movie *The Mirror Has Two Faces*. She plays an English Literature Professor at Columbia University. In the lecture scene, she is teaching class in a large lecture hall with over 300 students. Even though the instructional method is lecture with no slide show presentation, film, handouts, or visual aids, her teaching style is so engaging that the eyes of her students follow her around the room and hold onto her every word. She uses movement, hand gestures, real-life examples, and group discussion while she interacts with the crowd through natural facial expressions—smiling and laughing. She approaches her learners, injects appropriate humor, displays an inviting demeanor, and uses direct, open, and overhead questioning techniques. She is engaging, electrifying, and stimulating. She ends the class with a "teaser" so her students will return. She is wonderful. Her learning audience appears to be interested in the subject matter, excited, and engaged, which most often results in a positive learning outcome and a memorable training moment.

It's easy to remember the good teachers—from grammar school, high school, college, day camp, music class, or dance class. They are the ones you remember because they *created memorable training moments*. The lessons they taught are not forgotten.

Memorable Teachers and Trainers

My sisters and I still talk about Mrs. McGhee who was our elementary school tap dance teacher at the Shoesmith Park District—40 years later, we still talk about her and remember the dance moves she taught us and how exciting and engaging she was as a teacher. There was one dance routine that I still remember because we practiced for 10 straight days. Whenever I hear the song "The in Crowd" by the Ramsey Lewis Trio, a vision of Mrs. McGhee comes into my mind and I immediately think about her teaching methodology.

My mom was my first English teacher. There were many memorable teaching moments because she was always correcting me when I committed grammar abuse by using slang, informal colloquialisms, or ending sentences with prepositions. Often, she wouldn't acknowledge me if I spoke incorrectly, so it forced me to repeat the sentence—the next time correctly. I always hated it, but now I understand why she did it.

Memorable training moments also are associated with Clara Walton-Smith, my middle school typing teacher and Jan (John) Osada, my high school English teacher. My high school years were not always focused on school. At times, I was the rebellious teenager who didn't always do the right thing and preferred the social aspect of my high school years. For example, I was voted homecoming queen, prom queen, and most popular during this era. I really didn't enjoy school, until I attended college, but I loved my high school writing classes. I was always on time, sat right up in front, and always anticipated my next assignment from Mr. Osada.

Dr. Edward Homewood, my college English professor and Dr. Todd Hoover, my graduate school instructional design professor both brought me memorable moments. Dr. Lorraine Moline-Granieri, the Dean of Faculty at the first business school where I taught technology classes was passionate and had the greatest teaching style. She was similar to Barbra Streisand in the movie *The Mirror Has Two Faces*. She was a joy to watch—I wanted to be just like her.

Mary Stearns Sgarioto was my first editor who created many memorable training moments for me. I worked with a team of technical writers at a software company and she was the technical editor for 12 of us. I was new to writing this genre. There was a bit of a learning curve because technical documentation and writing instructions were something new. Instructional writing requires short sentences that are clear, concise, to the point, and easy to understand. Generally, people don't read technical documents for pleasure; they are looking for information to learn or quickly solve a problem.

We were the first generation of technical writers in the early 1980's. Prior to that, engineers, scientists, and other technical professionals were responsible for writing user guides, job aids, and training manuals. During this time technical writing wasn't even a college major. Many of us entered into this profession by accident because we were either Subject Matter Experts (SMEs), worked with computers, or showed an interest in writing. (At least this is how I started in the profession.)

As the computer industry blossomed in the 1980's, users began to complain that instructions were too technical, unclear, and complicated. The job of a technical writer became an economic necessity to reach a mass market of non-technical people. The computer industry needed individuals to translate technical data into a "user-friendly," non-intimidating language and to establish clear communication between the computer program and the person using the product. I still have the first edit Mary did for me in 1988. It was filled with red proofreader's marks and a request for a major rewrite. We didn't have computers and track changes back then. Instead, we wrote in longhand on yellow pads, gave it to the word processing department, proofed their draft, sent it back to them with revisions, and finally submitted the draft to the editor. This process would take days and several iterations. I learned so much after my first edit. Of course, I was a little shocked by all the edits, but all creative people know that you can only get better with honest feedback. She is still the person I call when I am stuck with grammar issues, word choice, and writer's block. She is also the editor of this book.

I really love what I do! My training career started with IBM two days after receiving my undergraduate business degree. I knew immediately that this would be my craft. Being part of the educational process, watching people learn, and helping employees perform their job tasks because of my involvement brings me great joy. I know that my experiences as a *Technical Writer* of computer learning materials, *Corporate Trainer* of software users, and *Instructional Designer* of curriculum, instruction, and training programs, contribute to successful learning. I also know this is a gift from GOD because I love it so much. *Psalms 16:9—True joy is deeper than happiness; we can feel joy in spite of our deepest trouble.*

Acknowledgements

I dedicate this book to several people. The first person is my husband Dr. Damon Arnold, MD, MPH. He practices internal and occupational medicine. As the chief physician for the state, he is responsible for population health. He has dedicated his life to medicine, public health, helping sick people, and making legislative and policy decisions that affect healthcare in this country. In addition, he is a decorated Colonel in the United States military and specializes in aviation medicine. Working as a Flight Surgeon, he completed two tours in Iraq and a variety of overseas medical missions. He has committed his life to public service and defending this country. He is the light of my life and the person for whom I love to begin and end each day.

Mary Stearns Sgarioto is a gifted writer, trainer, and technical editor. She has made me the writer I am today. She holds both a B.A. and an MFA (Master of Fine Arts) in English, and an M.Div (Master of Divinity) from the Lutheran School of Theology at Chicago. She is truly one of GOD's greatest creations put on this earth. I am so glad that I met her 23 years ago.

Lorraine Johnson Lynch and Stephanie Johnson Gaston, my two great supportive sisters (both work in public service), have been my backbone forever through every happy and sad moment of my life. Tommie and Nicole Lynch, who arrived in the 1980's, changed our lives and brought us joy by watching them develop into perfect children and fully functioning successful adults. Tyler and Mya Gaston, my new niece and nephew, give us the joy of watching their development, thanks to Rodney and Stephanie.

My mom, Pearline Haywood is highly intelligent and made sure my sisters and I were, too. She is organized, detailed and task-oriented, and socially conscious. When it was time to go to college, it wasn't "if" you are going—it was "where" are you going?" In the 1960's my sisters and I walked along with her in picket lines because she wanted equal opportunities and a better world for us, which we now have. My mom and two sisters are all smart and overachievers.

My dad, Clarence Johnson, has always been hard working and taught my sisters and me to be the same. He was an entrepreneur and business owner at the age of 25 and continued until he retired. He is highly analytical with the ability to recognize, analyze, organize, and solve issues and problems effectively. He embodies all of the success factors—handles conflict professionally, listens skillfully, works hard, is highly motivated to succeed, is independent, and highly structured. His wife Jackie Johnson is also highly skilled and intelligent. When our girlfriends got dolls for Christmas, my dad bought toy cash registers for my sisters and me so we could learn how to count and understand business—go figure, my oldest sister Lorraine has been employed as an accountant, auditor, controller, deputy city finance commissioner, chief financial officer, helps start up businesses, and teaches accounting principles. Her career began in the banking industry where she climbed the ladder to vice president. She left corporate America because she was recruited to lead the accounting department (as controller) for the largest human rights organization in the country.

My youngest sister Stephanie is highly analytical and worked as an aeronautical engineer for Boeing Corporation (manufacturer of satellites, commercial jetliners, and military aircraft). She always had a fascination with airplanes and was building and making them at the age of 13. She recently left corporate America to teach mathematics in the Dallas school system. She mastered every level of mathematics including discrete math, differential equations, and calculus 9. I'm not sure why I didn't get the math gene—math was always my worst subject. My dad has always been an inspiration because he had a great story to tell and is the reason I aspired to become a business woman, carry a briefcase, wear a blue suit, work in corporate America, and finally become a corporate entrepreneur.

Friends warm my life, comfort, support, and applaud me. I wouldn't be who I am without my sisters and a few friends, so I need to acknowledge three great friends who have been through every issue, triumph, and struggle over the journey of my life: Lurie Moran-Brown, Carol Proctor, and my first cousin, Elizabeth Johnson-McGhee.

Contributors

A special thanks to my content expert, Valerie Washington, M.S.Ed. She has over 20-years of expertise in the field of Instructional Technology and Human Performance Improvement. As founder of *"Think 6 Results,"* a model for continuous learning, her goal is to engage minds and create continuous learning that closes the strategy-to-performance gap.

A special thanks to my content expert, Dr. Stacy Saxon, an educational psychologist who is one of the most gifted trainers I have had the pleasure to work with.

A special thanks to my content expert, Beverly Rico, with an M.S. in Training and Development. As the receiver of the ICEOP scholarship from Roosevelt University, AT&T Circle of Excellence Award and the winner of the "Above and Beyond " and "Way to Go" awards, she offers 20 years of experience in the field of training, instructional design, facilitation, project management, and strategic planning.

A special thanks to my editor, Mary Stearns Sgarioto, my grammar and writing rescuer.

A special thanks to a sincere colleague Rhea Steele, B.S. As President and CEO of Edge Technological Resources, her firm offers management, education, and technology consulting services. Rhea and I are former IBMers who decided to pursue business ownership. She has allowed my firm TechnoWrite, Inc. to participate in many projects which have shaped me and allowed me to blossom into the professional I am today. I owe an enormous amount of gratitude and appreciation to her.

I've had the pleasure to work in an industry that I love. I have also had the privilege to work with some outstanding training professionals and colleagues over the past 25 years. These are individuals who I have watched deliver excellent training and were extremely impressive. Each of them possess the attributes that I feel make outstanding Trainers.

These attributes include capturing and keeping an audience, showing care and concern for their learners, having a willingness to go the extra mile (stay late, miss badly needed breaks, or call students on their own time) to make sure learners understand the content, creating positive learning outcomes, and demonstrating a passion for the craft of a Trainer. You can read about these individuals in Chapter 9. They include Charlotte Cager, Adel Etayem, Ellen Lehnert, Thelma Reed, Beverly Rico, Mary Sgarioto, Luisa Vercillo, and Candace Zacher.

Understanding the Role of Instruction

Although the focus of this book is on *Techniques for Technology Training*, if you are new to the training profession, let's take a few minutes to *briefly* review the phases of instruction and the importance of the Instructional Designer's (ID) role.

What Is Training?

The words training, instructing, teaching, and coaching are often used interchangeably. Training refers to acquiring skills, knowledge, and competencies needed to perform a task. Although you typically learn from instruction—you can also learn from watching others, life experiences, and unconscious behaviors picked up from someone else. Education is broader and affects the mind in a holistic way while training is often quick, fast, and to the point.

Since my expertise is software implementations and deployments, my job is to teach people to use software programs that provide specific skills required to perform job functions. My goal is to make sure participants leave my training workshop and return to their jobs with the ability to perform the tasks learned in training. For example, if I'm conducting a training workshop for a new purchase order system, my final question for the learner will be, *"Can you return to your desk and create a purchase order using the new system?"* In corporate education it is often referred to as just-in-time training, accelerated learning, or rapid training development. I often work with organizations that need to learn new skills quickly. In today's world there is software all around us, and the delivery of training needed to learn it must be clear, concise, and easy to understand while minimizing learner frustration.

Training in the workplace may be necessary to upgrade and update skills or improve job performance. Sometimes this is self imposed or required to maintain or keep a current job. For example, during the early 1990's when organizations were moving away from manual processes and huge mainframes with batch processes, learning new computer skills was mandatory. Personal computers evolved in the workplace around 1990, quickly replacing electric typewriters, word processing departments, and administrative assistants. Now, computers are more compact (laptops, personal digital assistants, notebooks, phones and smart phones), faster, and easy to use. In summary, training and professional development are necessities throughout a person's work life.

How Does Instruction Begin?

Every learning outcome must begin with a well-thought-out design. An Instructional Systems Design (ISD) [1] model is a methodology and approach used to design training programs. It is the core of good instruction and allows you to create the training plan from the inception (the idea) through completion (training delivery).

The U. S. military in the 1940's were the first to use an ISD model for the development of their programs. As the model matured, a variety of additions and variations were created such as Ropes; Dick and Carey; Analysis, Design, Development, Implementation, and Evaluation (ADDIE); Kemp; Morrison; and the Ross Model, to name just a few. However, almost all training design is based on the generic ADDIE model.[2] This five-phase approach includes analysis, design, development, implementation, and evaluation. Each phase has an outcome that feeds

---

1   Hodell Chuck, *ISD From the Ground Up*, American Society for Training & Development, 2000

2   Piskurich George, *Rapid Instructional Design*, John Wiley & Sons, 2006

into the next step in the sequence. In the *analysis* phase, designers gather details about the learning audience and their needs. The *design* phase is used to determine learning goals, objectives, and training strategy. The *development* phase involves designing and writing the learning materials, which includes job aids, manuals, e-learning, and self-study. *Implementation* is the delivery of the training to the learning population. Finally, *evaluation* measures the merit and worth of the training program by evaluating the learner's reaction, learning, behavior, and results. The five-step ADDIE model is briefly described below.

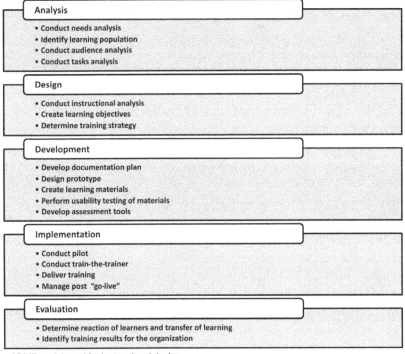

Analysis
- Conduct needs analysis
- Identify learning population
- Conduct audience analysis
- Conduct tasks analysis

Design
- Conduct instructional analysis
- Create learning objectives
- Determine training strategy

Development
- Develop documentation plan
- Design prototype
- Create learning materials
- Perform usability testing of materials
- Develop assessment tools

Implementation
- Conduct pilot
- Conduct train-the-trainer
- Deliver training
- Manage post "go-live"

Evaluation
- Determine reaction of learners and transfer of learning
- Identify training results for the organization

ADDIE model used for instructional design

Trainers and Instructional Designers—Making the Connection Together

In my 25 years in this industry, I have often witnessed a "disconnect" between IDs responsible for the design and development of the instruction, and the Trainer who is responsible for the delivery. There must be collaboration and understanding between these two roles. This disconnect can often lead to a breakdown in the overall goal of the training and/or the organization. Even if a Trainer does not wear multiple hats, it is important to have a basic understanding of both roles.

The Role of the Trainer

Several years ago, I evaluated a seasoned payroll manager who was conducting training for a group of new payroll clerks for an online system. She was the Subject Matter Expert (SME) recently promoted as the department Trainer. She didn't follow the training manual designed by the ID. Instead, she decided to teach the challenging topics first. The class began with an intimidating scheduling screen that was used to enter payroll schedules and dates. The screen was multifaceted and displayed vertical rows of numbers, an assortment of fields, and five different color-coded sub-menus. It was a complex screen even for a super-user. An illustration of the screen's components coupled with callouts and explanations should have been reviewed first. The Trainer did not review basic payroll concepts and failed to ensure that the prerequisite topics were fully understood before teaching the scheduling topic (which was identified in the training manual by the ID). Needless to say, no one learned anything on that day. After class, she didn't understand why students were confused. When I asked her why she started with the difficult topics first, she responded, *"If they get the hard topics first, the rest will be easy."* Most professional Trainers know this is absolutely mistake number one because you should never set up your students for failure. You should always teach the simple topics first and allow learners to build on what they already know.

I have also seen Trainers who won't use all of the learning materials the way they were designed because they didn't fully understand basic ID concepts, such as: teaching from simple to complex, understanding the prerequisite needs of the audience, aligning the learning objectives with company goals, using various questioning techniques and assessments that match the objectives, and relating system functions to job tasks.

Even if you are a skilled Trainer with effective delivery skills, it is important to understand the process and methodology needed to design good instruction. There may be instances when it is necessary to modify the curriculum topics. However, keep in mind that a lot of research, analysis, and review of the content were conducted before the Trainer is in possession of the learning materials.

The Role of the ID

Likewise, it's important for IDs to understand how the training delivery alone can have an impact on the learner. I have worked with IDs who have never conducted instructor-led training and don't fully understand what it is like to stand up and teach content from a training manual. I have worked in environments where the ID will design the course and hand it off to the Trainer only to discover that the Trainer didn't use all of the material as intended.

IDs must understand that there may be deviations in the flow and content due to certain circumstances, and that Trainers must always be prepared to deal with and respond to the unexpected. Some examples of these situations include:

- Additional topics are needed

- Assessments don't work well for a particular group

- Topic needs further explanation than what was provided

- Mixed skill levels exist in the classroom

- Adjustment to the topics is quickly needed to meet the needs of the audience.

There are numerous situations when a Trainer must make a quick decision about modifying the design of the instruction.

I began my career as a Trainer of software users. Four years later, I became a Technical Writer. Six years later I returned to graduate school to learn instructional design. I always believed I was an effective Trainer. After all, the ratings at my end of class evaluations were always superb. However, once I fully understood the ID process, I made an even bigger impact on my learners. It is important for both Trainer and ID to understand each other's role.

As stated earlier, Trainers need a basic knowledge of how instruction is designed. The remainder of this chapter will describe the phases of instruction and the importance of why IDs and Trainers must work together.

Identifying Onboarding Tasks for IDs and Trainers

Once the ID is given the job of designing the instruction, there are a few tasks that must be performed before the design process begins. It's also a good idea for Trainers to perform onboarding tasks to learn as much as possible about the company, project, and people they will work with.

First, the ID and Trainer must learn as much as possible about the current state of the existing business processes. It is important for IDs and Trainers to have a general understanding about what learners do in their current work environments in order to create relevant, usable materials.

Next, the ID and Trainer must learn about the new system's functionality. This can be accomplished by gathering and reading resource materials, such as: future state, vision, process flows, or functional design documents. This step also involves conducting interviews with individuals involved in the system's development and is mandatory. For example, interviews should be conducted with system designers, system architects, programmers, management, SMEs, and the user community (if applicable).

Finally, if you are delivering training for an industry for which you have never worked, conduct your own research to learn about the background and details about the company's products and services.

Using an Instructional Design Model

Phase 1—Analysis

The training intervention typically begins with a *needs analysis* to justify why it is necessary to spend training dollars or make an investment in human capital. Once the need is established, the analysis phase begins with identifying key characteristics of the learning population.

Conducting the Audience Analysis

Before training can begin on the new system, specific skills and concepts must be mastered before learners can use the new software. An *audience profile* will be conducted to identify traits of the learning population such as training experience, education, computer literacy skills, typing skills, desktop skills, and training preferences.

A *skills discovery* or *pre-test* will identify whether the necessary prerequisite skills required for the new content exist. Successful training implementations occur when they are designed to match the skill levels of the participants. If prerequisite computer skills, concepts, and certain entry-level skills have *not* been mastered, there will be a mismatch between instruction and the abilities of the learners to perform the tasks. The mismatch of skills in the classroom and the re-training of employees can be avoided by conducting an audience analysis. This step will also address how to handle the various user levels (novice, beginner, intermediate, or advanced) that exist and how to manage the integration of varying skill levels in the classroom. The results of this analysis will conclude with a description of a learning gap that exists between "what is" and "what should be."

Conducting the Task Analysis

After you have identified "who" will use the training, let's find out "what" job tasks they perform. Adults need training that is relevant to their current job tasks. A *task analysis* identifies each job task that is performed by the learner and its relationship to the new system. Avoid developing training and documentation around software functions. Instead, training must be related to job tasks and responsibilities, *not* product or system features because technology is constantly changing. For example, perhaps a new accounting software system offers 30 newly enhanced features. However, the payroll clerk will only use 10. Although it would be nice to know the functionality of all 30 features, the instruction for the payroll clerk should only focus on the 10 that are relevant to his or her job.

The task analysis also allows the ID to create comparison learning scenarios for training. For example, *"This is how you created a payroll report in the old system; this is how you will do it in the new system."* Often, learners will ask the following questions: *"What does this have to do with my job?"* or *"We did it like this in the old system?"* It is important for Trainers to learn as much as possible about the job-related tasks to help achieve instructional goals.

Phase 2—Design of Training

After the training need has been identified, using the data collected in Phase 1 (Analysis), you are now ready to design training. Although each organization uses its own methodology, I begin by asking a few questions:

- What tasks can the learners currently perform?

- What does the organization want them to be able to do?

- What are the knowledge, skills, and abilities (KSAs) that must be developed?

- What is the new curriculum?

- What is the strategy and approach?

- What tools are needed?

Performing the Instructional Analysis

The design phase begins by making decisions about the curriculum. An instructional analysis determines how to design the curriculum— what content to present and when to present it. IDs often build *learning hierarchies* to illustrate the sequencing of learning events and to identify any unnecessary instruction.

Curriculum mapping documents the content that will be taught. This includes how many curricula, learning modules, and courses are needed. There could be one or more curricula for a given training implementation. For example, a client might design curriculum 1 for industry concepts, curriculum 2 for system tools, and curriculum 3 for job tasks. The task analysis (discussed earlier) results will reveal how to organize the job tasks into learning components. Each curriculum will have as its own set of learning modules and courses.

The *training goals*, *training purpose*, and *training scope* are also defined for the curriculum. Details about each course will be provided using the data described in the table below.

Course Name	Course Description	Learning Objectives	Audience	Departments/ Job Titles	Assigned Tasks

Curriculum map example

A *design document* will be developed for each training course. It serves as a blueprint for building the course during the development of the materials. The design document includes a detailed course outline (chapters, topics, and subtopics), learning objectives, instructional methods, activities, exercises, and assessment tools. It can also be used to connect to assessment data and instructional strategies.

Developing the Learning Objectives

As the Trainer, ask yourself: *"What do the learners need to know?"* and *"What do I want to teach them?"* A *learning objective* identifies what the learner will be able to do at the end of each instructional unit. A learning objective is observable and measurable. For example:

> At the completion of this workshop, you will be able to enter a purchase order using the Lynch payroll system with 100% accuracy.

Learning objectives are sometimes overlooked but are extremely important for training! They determine and measure the results of the training intervention. A learning objective tells the Trainer what will be taught and tells the learner what he or she can expect to learn. Learning objectives have three components: the performance, the condition of performance, and the criteria.

1. The performance (observable and measurable) includes what is required. What will the learner be expected to do after completing the instruction?

2. The condition of the performance includes the tools, methods, job aids, equipment, etc. that are needed. What are the conditions under which the behavior will be expected to occur?

3. The criteria describe the expected quantity or quality. It is also observable and measurable. How is the learner expected to perform? For example, with 100% accuracy, with no errors, within 15 minutes, or with 90% accuracy.

Let's look at the objective again.

> At the completion of this workshop, you will be able to enter a purchase order (performance) using the Lynch payroll system (condition), with 100% accuracy (criteria).

Learning objectives are written using measurable or observable action verbs. In 1956 educational psychologist Benjamin Bloom developed *Bloom's Taxonomy of Learning Domains*[3] which concluded that there are three domains of learning such as cognitive (mental skills or knowledge), affective (attitude, feelings, and emotions), and psychomotor (manual or physical skills).

In the cognitive domain, there are six categories that move from the simplest of concepts to the more complex. They include knowledge (level 1), comprehension (level 2), analysis (level 3), application (level 4), synthesis (level 5), and evaluation (level 6). Each level must be mastered before the next one can be achieved. Learning objectives are written based on this premise.

---

3   Bloom Benjamin, *Taxonomy of Educational Objectives Handbook 1 Cognitive Domain*, Longman, 1984

The list below provides a *few* examples of verbs that can be used when designing and writing learning objectives.

Knowledge	Comprehension	Application	Analysis	Synthesis	Evaluation
List	Translate	Demonstrate	Prioritize	Develop	Translate
Identify	Interpret	Illustrate	Verify	Construct	Test
State	Estimate	Perform	Differentiate	Create	Measure
Arrange	Contrast	Translate	Categorize	Specify	Revise
Define	Sort	Complete	Examine	Assemble	Estimate

Bloom's cognitive domain hierarchies

Using the Learning Objective to Validate Learning

Developing learning objectives is an important instructional strategy because they can be used to assess and measure learning outcomes. Let's look at another example.

> At the end of this chapter, you will be able to search for a purchase order, using basic Boolean search operators, with 100% accuracy.

After the instruction, you then can assess (discussed later) your learners on the learning objective. To determine learning outcomes, you can use a variety of instruments such as tests, questioning techniques, and demonstration. Whatever instruments you use, make sure you design assessments that relate back to the learning objectives. If you haven't taught them, how can you test them? Do not test someone on a topic that you haven't taught them. Using the learning objective above, you could assess the learner by asking him or her to:

- Demonstrate how to search for a purchase order.

- Explain the steps to search for a purchase order.

- Complete a test question on this topic.

What Is the Training Strategy?

Now that you have completed the onboarding tasks, described the audience characteristics, identified the job tasks of the learners, and defined the objectives to be achieved, it's now time to determine the strategy and approach. Will the instructional approach for this training intervention be instructor-led training (ILT), web-based training (WBT), virtual (asynchronous or synchronous) delivery, self-study, or blended? Let's look at an example of how an organization chose a blended approach as a training strategy.

• Training Strategy 1—The Lynch Corporation will use an *ILT training delivery* consisting of 12-15 learners per classroom. The instruction will be accompanied by a student guide, job aids, and a PowerPoint presentation to serve as the instructor's guide. The classroom will use a training database complete with simulated data of the new environment to allow learners to role-play realistic business scenarios. A *skills practice lab* will be available with practice scenarios for learners to perform self-paced independent study.

• Training Strategy 2—The Lynch Corporation will use two methods for *WBT training delivery*. *Training videos* primarily used during the pre-go live phase will be created to describe basic system concepts. The 10-minute video provides a linear approach for learners to review details about the new system at any time (during lunch or breaks). In addition, *e-learning modules* will be designed using an interactive software program to capture and record screen activity. The program will demonstrate a procedure complete with mouse movement and text descriptions simulated for simple transactions. An interactive simulation with prompts and feedback will allow learners to practice the recorded steps.

• Training Strategy 3—The Lynch Corporation will also provide a *virtual training delivery* after the system is in production. A virtual synchronous classroom allows learners to participate in a live, instructor-led class. Weekly classes will be offered to provide complex problem-solving and troubleshooting techniques. The client chose a synchronous classroom to accommodate people from various time zones. Whiteboards, application software, audio conferencing, video conferencing, and chat rooms are available tools for this method.

Phase 3—Development of Learning Materials

After phases 1 and 2 are completed, an ID will be given the design document to use as a basis for developing and writing the materials. Instruction must be accompanied by quality documentation. This can include participant training manuals, instructor guides, job aids, e-learning modules, visual aids, and training videos. In most instances, the learning materials are used as a reference during instruction.

Developing the Learning Materials

As discussed earlier, all good instruction begins with sound instructional design principles using an ISD model. Although instruction is important, creating high quality, usable materials plays a significant role because they can be used as a reference once the training event is over.

In the development phase, the production of the learning materials occurs. IDs and/or Technical Writers should be skilled professional writers and editors. Their job is to look at the new system from the viewpoint of the user and act as a liaison between the programmer and the person using the new computer system.

Learning materials are designed to make a task, process, or procedure easy to perform while following the step-by-step instructions as a guide. Therefore, the learning materials should be clear, concise, and easy to understand. I often think of software users as "innocent" because they have been forced to learn new technology fairly quickly. The materials—whether user guides, job aids, online help, e-learning modules, or training manuals—should help to minimize the frustration of learning the new information.

Instruction manuals are not typically selected for pleasure reading. The average person does not want to spend a great deal of time reading and searching for information. For example, let's say that you are using a payroll system to perform a manual adjustment and suddenly you can't figure out which function to use. You open the training manual to the correct section to locate a quick answer to your problem. However, the step-by-step instructions are not clear. So now the training manual will be placed on a book shelf and remain there forever because it was unreliable.

The goal of IDs and/or Technical Writers is to create documentation that is useable, reliable, and true. That means translating technical concepts into user-friendly, non-intimidating language.

A *documentation plan* will be created. This includes planning the documentation project, providing deliverable dates, designing the page layout and format, working with SMEs to build the content, writing, editing, identifying standards and conventions, performing usability testing on the materials, conducting a review meeting to inspect the document, approving the materials, and the final production. The documentation plan coupled with the design document and other components identified in phase 2 (design), will serve as the blueprint for the development of the learning materials.

After the documentation plan is complete, a *prototype* is the first work product to be created. It provides the client with a preview of the look and feel of the materials. The prototype provides a layout of the page's design. This includes the format, content, and layout specifications (typography, heading levels, headers, footers, white space, conventions used, illustrations, and tables). It's a good idea to have your client sign-off on the prototype before committing to writing.

What about Assessments?

The ID will create various forms of assessments for the Trainer to ensure the learning objectives have been achieved. Competency checks will occur throughout the learning event using a series of questioning techniques (closed, open, reflexive, leading, and next-step), group discussion, or proof of performance via demonstration of a task. In addition, case-based and what-if scenarios will be used to problem-solve real-life simulations of the environment.

Phase 4—Implementation

There are four phases of the implementation process. They include pre-go live, pilot, classroom training delivery, and post-go live.

Pre-go Live Phase

There are a number of activities that can occur during the pre-go live period. The goal is to achieve user "buy-in" of the new system. This can include lunch and learn sessions, live demonstrations in designated rooms, posters, and newsletters to announce the new system, E-blasts, announcements on the organization's intranet, and system kick-off events.

Conducting the Training Pilot

It's a good idea to conduct a pilot class prior to the actual delivery to the learners. This will ensure that the courseware and Trainers are effective and that the facilities are adequate. After the pilot, a debriefing session is scheduled with the participants. Every aspect of the pilot will be evaluated and the following questions will be asked:

- Were there any system or data issues?

- Did the system's functionality work properly?

- Did the learning materials work?

- Was the training effective?

Next, any feedback can be incorporated before the actual training begins.

Conducting Instructor-led Training

Now you are ready to install the instruction in the real world. This phase includes the delivery of training and placing the training plan in operation. The Trainer must determine the best delivery approach and identify the *instructional methods* to be used to teach the class such as: hands-on, brainstorming sessions, group discussion, role play, demonstration, programmed instruction, walkthrough, team tasks, games, simulation, or exercises. Using a variety of instructional methods and classroom activities ensures learning will take place, reduces boredom, and keeps learners interested and excited.

Post-go Live Phase

On the day the system goes live, a drop-in training room could be set up to support the users. The room will have live PCs, and Trainers, SMEs, and system programmers will be available for consultation. One-on-one training, answering questions, and troubleshooting issues can be performed. The drop-in training room can be set up for a two-to-three week period after the conversion. *Floor support* could be offered to each department during the transition period.

Phase 5—Evaluation

Evaluation measures the effectiveness of the training program and determines the impact of the instruction. Evaluation tells us whether or not the training instructional goals were met. For example, *"Was there a transfer of learning?" "Were the learning objectives achieved?" "Can learners perform the tasks learned during training?"* or *"Did learning barriers exist?"* The *Kirkpatrick Model*[4] is a four-level evaluation model used to assess the effectiveness of a training program discussed later in Chapter 8.

Evaluating the Training Intervention

Evaluation is used to assess student achievement, evaluate the curriculum, and improve the learning materials. Evaluation provides the following benefits by allowing:

- *Learners* to express their feelings, attitudes, and reactions about the training they received.

- *Trainers* to receive feedback about the course content, trainer delivery, and facilities.

- *Organizations* to obtain evidence to support the implementation, justify the program dollars, and determine whether or not a return on their training investment was worthwhile.

In addition, *summative* and *formative evaluation*[5] may be conducted. A summative evaluation is typically conducted at the end of the event to provide judgments about the program's worth and merit. Likewise, a formative evaluation occurs during the implementation phase to identify

---

4   Kirkpatrick Donald and Kirkpatrick James, *Evaluating Training Programs: The Four Levels, Third Edition*, Berrett-Koehler Publishing, 2006

5   Rothwell William and H.C. Kazanas, *Mastering the Instructional Design Process*, San Francisco: Jossey-Bass, 1992

whether the program needs improvement. This allows the ID and Trainer to adjust and revise the instruction immediately. Evaluation tools are useful because mistakes can be corrected before they cause serious problems for the client.

Now that all phases of development have been reviewed, let's review the following case scenario and the final outcome of a training implementation that went badly. When a systematic approach is *not* used, it will often result in a failed deployment.

A Case Study: A Training Intervention Gone Bad

"What Can We Learn From the Past?"

After the completion of an intense 12-week training program, top executives asked the following question, *"Why aren't they performing?"* I was part of a group of *performance consultants* brought in to identify why training didn't work and what would be required for retraining. With the *Kirkpatrick Model*[6] levels one and two in hand, we began the preparation for our evaluation.

First, we began our discovery and research with focus groups. However, because the focus groups included supervisors, we didn't get a "true" picture from the employees. Next, we issued written surveys, which provided great insight. However, we received less than a 25% response rate. (A 50% return is the minimum threshold required for a valid statistical result.) Finally, we decided to conduct one-on-one interviews. Employees interviewed were frustrated, used colorful vocabularies, and one employee (who had 20 years with the company) shed a few tears. The employees clearly wanted to vent their frustrations to anyone who would listen.

What Was the Purpose of the Training Initiative?

A vigorous, intense 12-week training program was implemented in two parts. Part one involved learning about the industry. Most new hires had

---

6   Kirkpatrick Donald and Kirkpatrick James, *Evaluating Training Programs: The Four Levels* 3rd *Edition*, Berrett-Koehler Publishing, 2006

never worked in a customer service environment and were new to the industry. In addition, in order to function in the new job, a state license was required. Employees attended daily training classes for four weeks to learn about the industry. To help them prepare and pass the state exam, they were given the option of attending class after work or taking a self-study course. During this four week period, employees simultaneously attended daily training classes and prepared for the license exam.

Part two involved learning eight different proprietary software programs specific to the company. In addition to understanding a new industry, employees also needed to learn about company policies and procedures, the product line they would support in the customer service environment, and how to function on the job. Six-hour training classes held four days per week were conducted over a period of eight weeks. Qualified professional Trainers who were skilled in adult learning conducted instructor-led training. They were also subject matter industry experts with extensive product knowledge.

What Did We Learn During the Evaluation?

Our findings concluded the following:

1) Instructor-led training was conducted using three instructional methods (hands-on, simulation, and lecture):

- A vendor-supplied test database was used for learners to practice hands-on classroom exercises for a few courses. However, the simulated database used data from a totally different industry from the one in which the learners worked. The data used was not unique or customized for the company.

- The lecture was only accompanied by a PowerPoint presentation. Even though the presentation provided details about the software functions and showed copies of the system screens, learners did not have a true picture of the system because there was no interaction with the software. Learners basically watched the screen and took notes.

- One crucial course allowed learners to listen to "live" phone calls in a customer service environment. However, the calls they listened to were for a manufacturing company, not their organization. Students didn't learn about their customer base or what to expect on the job. The calls were not unique or customized for the company.

2) No target audience profile was conducted prior to training. Therefore, prerequisite skills were not identified. Employees attending the class had a mixture of PC skill levels from novice to expert. Many of the seasoned employees were well versed in the industry and organization but were PC novices. Also, because of their lack of computer skills, they were uncomfortable and intimidated by the super users. Therefore, they didn't fully participate in the class because they felt intimidated. The PC novices desperately needed basic computer literacy classes such as *Introduction to PCs, Using the Mouse, Computer Terminology, and Introduction to Windows*. When we interviewed the PC experts, they told us that the entry level content that was provided was boring. This mismatch of skills in the classroom and the lack of prerequisite knowledge was a challenge to the Trainer and all participants. This contributed to the lack of successful training outcomes.

3) The Trainers were experts and seasoned professionals who knew the company and industry well. Many of the new employees hired were Generation Xers complete with sassy and creative computer skills. When we interviewed the Trainers, they had issues relating to the Xers. The cultural clash caused Trainers to constantly tell the Xers *not* to talk on the phone, *not* to surf the web, *not* to use text messaging during class, and to return from lunch and breaks on time. These constant interruptions were frustrating to Trainers and resulted in a failed knowledge transfer. A few Trainers asked for a class in generational training.

4) After months of training, learners returned to their desks to practice the newly learned programs. However, no Structured-on-the-Job-Training (SOJT) was available, and there weren't any mentors or coaches available either.

5) We reviewed the training manuals to determine: a) if the learning objectives were derived from job performance; b) if the learning materials were task-centered; and c) if the end-of-unit assessments matched the skills specified in the objectives. (The answer to these three questions was no.)

On the day the system went live and the newly trained employees moved to their desks to begin answering phone calls, they couldn't help any customers. On the third day, senior management asked, *"Why aren't they performing?"*

Can Anyone See the Obvious Gaps?

To create a sound program and avoid training failures, the ID and Trainer should fully understand the needs of the learner. Creating training based on perceived skills without a thorough understanding of the learner might not result in a successful outcome. The next chapter focuses on understanding the role of the learner.

Understanding the Role of the Learner

What Is the Role of the Learner?

Before you can make your training events memorable for your learners, you must *first* understand the role of the learner. Placing people in learning mode removes them from their normal routines, and they are faced with something new or different. This "change" may make some individuals vulnerable, stressed, challenged, or even afraid. It is important to set up conditions for successful learning.

Classrooms can create anxiety for some people—even the most intelligent. Many adults may not have been in a classroom for years and may have struggled academically in the recent or distant past. A safe, comfortable environment must be established for learning to take place and 90% of that learning environment is a reflection of the Trainer's attitude and frame of mind. If you want learners to be eager to learn and use new skills, fear must be minimized. Keep in mind that learners might:

- Dread speaking in front of a group

- Fear making a mistake (their egos are on the line)

- Feel more comfortable sitting and observing.

Be sure to generate a safe, open, non-judgmental, non-competitive atmosphere where learners feel respected and their opinions are valued. This is the only way to get participation in the classroom. Treat contributions and responses as opportunities for showing approval.

Identifying Learning Behavior

Effective training involves successfully applying training methods to achieve intended learning goals. The methods you use to achieve successful learning depend on who receives the training, the knowledge and experience of the participants, and their personal motives.

Who Are Adult Learners?

Pedagogy (the science of teaching), is an excellent model for education. Pedagogy is teacher-focused because the teacher makes the decisions about what will be learned, how it will be learned, and when it will be learned. In the pedagogy model, teachers direct learning, and it is often used to teach children.

Andragogy (the science of teaching adults), emphasizes that adults are self-directed and expect to take responsibility for their decisions and education. Malcolm Knowles[7] is one of the innovators of adult learning theory from the early 1970's. Andragogy assumes that adult learners are:

- Self-directed and must be free to direct their individual learning

- Want to know the reason for the new learning

- Need to learn experientially

- Approach learning as problem-solving

- Learn best when the topic is of immediate value

- Experience a decrease in vision and hearing ability, an increase in long-term memory, and a decrease in short-term memory

- Require a variety of stimuli and training methods

---

7   Knowles Malcolm, Holton III Elwood F., Swanson Richard, *The Adult Learner*, Butterwort-Heinemann Publications, 1998

- Enjoy solving real-life problems and prefer hands-on methods of learning.

What are the Reasons and Motives for Adults to Learn?

Have you ever had a group of disgruntled employees come into your classroom? The first thing they will ask is, *"Why do I need to learn this?"* Clearly, the reasons and motives for learning are different for adults than children. While parents and grades motivate children, adults need different stimuli. In practical terms, instruction for adults must focus more on the process and less on the content being taught. As you train adult learners, please keep in mind that they:

- Bring a tremendous amount of life experience into the classroom, handle the day-to-day demands of life, have busy schedules, and have constant life changes

- Want task-oriented materials organized by tasks, not subjects; adults are most interested in learning subjects that have immediate relevance to their jobs or personal lives

- Need to see the value and benefit in what they are learning

- Are motivated to learn for real life benefits and personal advancement such as better wages, recognition by peers, promotion, public displays of achievement (placards, certificates, company newsletter), or enhanced job skills

- Are self-directed; and therefore, instruction should allow learners to discover things for themselves

- Find serving humankind motivating and enjoy giving back, serving the community, building relationships and friendships, or relieving boredom from routine events

As a Trainer, address these factors by explaining what participants will learn, why the new information is important, and the benefits of the learning outcome.

What Behaviors Can Influence Learning?

When training adults, be aware of the psychological and physiological factors that may affect adult learning behavior as described below.

- Change can be an obstacle to new learning. Any type of new direction may cause fear and anxiety. Organizations can use change management strategies to help learners get ready for the change such as brief overview sessions, lunch and learn events, short training videos, kick-off events, and drop-in sessions. There are a myriad of ways to transition to the new learning initiative.

- Previous learning experiences and preconceived notions from the past can hinder performance if the learner feels threatened by the training event. For example, if a person has had negative learning experiences or lacks confidence in his or her abilities, this most often can affect learning.

- Breaking old habits can interfere with doing things in a new and different way. For example, dispensing with the *"We've always done it that way"* attitude can have a major impact on helping resistant learners make the transition.

- Be aware of a learner's stage of development. Depending on the stage of the learner's development, there may be different learning motives and expectations. For example, there are six major stages of development:

  o 18-23 – entering early adulthood

  o 23-28 – early adulthood

  o 29-35 – adulthood

  o 35-42 – entering middle adulthood

  o 45-55 – middle adulthood

  o 55-65 – entering late adulthood

What about Generational Training?

Another important aspect of understanding the role of the learner is becoming aware of generations colliding with each other. How will you deal with a class comprised of traditionalists, baby boomers, Generation Xers, digital Yers, and millennials in your classroom? As discussed earlier, each learner's stage of development will influence his or her motivation levels. It is important to understand the differences that can affect your training events. For example, you might teach a retirement class for a group of 18-23 year olds differently than you would for a group of 45-55 year olds. Likewise, the motivation and interest level will be different for each group.

Understanding the role of the learner is the prerequisite for designing and implementing successful training. The next chapter focuses on understanding your role as the Trainer.

Understanding Your Role as the Trainer

Creating an Environment for Success

Before you can make your training events memorable for your learners, you must *first* understand your role as the Trainer. A safe environment that is conducive to learning is important for learner success. A cohesive environment begins with a qualified Trainer who is also a SME who makes learners feel welcomed. It's also important to understand learning styles, processing differences, and possible barriers that might affect outcomes. In addition, because an atmosphere of success means everyone will feel welcomed, you must manage diversity in the classroom as well as your own preconceived notions.

Identifying Audience Characteristics

It is beneficial to learn as much as you can about the people you will be training. Your learners are your customers and they deserve the best value for their training dollars. An *audience analysis* allows you to obtain a complete profile of the target population. Several characteristics are evaluated such as the learner's education, experience in present or related jobs, job performance requirements, language or cultural differences, physical or psychological characteristics, interests or biases, training background, length of time on the job, training preferences, motivation levels, and computer skill levels (novice, beginner, intermediate, advanced), to name a few.

Conducting an audience profile identifies specific details about the learner's skills and abilities. This information will help you create a more successful learning environment and set clear expectations. A Trainer without an audience profile will have no valid information about the learning audience. This can result in preconceived notions, incorrect assumptions, and biased opinions about a specific population.

What Are Prerequisite Skills?

Prerequisites are the skills required *before* the new information can be learned. For example, if you wanted to teach four-year-old Jacob how to tie a shoe lace, what does he need to know first? There are two prerequisites skills required before the actual instruction can begin. First, he needs to know his left hand from his right hand. Then, he must know how to tie a knot. Likewise, an employee who is learning to use a new computer system must first know how to use a mouse to navigate.

If learners do not possess the necessary prerequisite skills, they will not be successful and the training intervention may fail. Often, many of the barriers faced by learners exist because they do not possess the prerequisite skills.

How Are Prerequisite Skills Determined?

The beginning level of instruction is the course the learners will take when they walk into your classroom. For example, if you are teaching the course "Internet Level 2" you want your learners ready to take this class when they arrive for training. There are a few assumptions that can be made about the individuals who signed up for this course. For example, because this is a level two class, you can assume they have basic computer skills and know how to navigate the Internet. However, what happens if you are ready to teach, assume they are ready, and then determine that the individuals who signed up for the class don't understand basic concepts about the Internet or they lack basic computer skills.

One method for determining prerequisite skills is to administer a pre-test. A pre-test is designed to measure those skills that have been identified as critical to beginning instruction. It also determines skill levels required for using the new system. If learners lack these prerequisite skills, they may experience difficulty with the instruction. A pre-test is used to:

- Measure entry (prerequisite) skills

- Obtain evidence and proof of skill levels for all participants

- Avoid subjective opinions by managers and supervisors

- Identify and measure the critical skills that learners need before they begin instruction on the new system.

Let's look at an example of how identifying prerequisite skills saved the organization training dollars, reduced retraining, and achieved a return-on-investment (ROI).

The Lynch Corporation is scheduled to deploy a new online financial management system. A total of eight courses will be offered as part of this implementation. Computer skills are required for all the courses because the new users will add, edit, delete, and view customer details. The first course is called "Creating a Purchase Order." This is the beginning level of instruction, so learners should have the necessary prerequisite skills to take this course when they arrive. Instead of making assumptions, the Training Manager conducts a computer literacy pre-test to see which employees are ready to enroll in the course. The pre-test will allow the manager to identify basic PC and Windows skills and then determine the levels (novice, beginner, intermediate, or advanced) of the workers. The learning population consists of 295 employees. Let's review the results.

The illustration below provides the results of the pre-test.

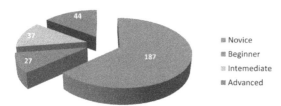

What do the results tell you about this population?

Is this organization, how many learners are ready to begin training on the "Creating a Purchase Order" course?

In the above illustration there are 187 individuals who are <u>not</u> currently ready to take the first course. This means a few prerequisite courses must be completed before they can take the "Creating a Purchase Order" class. Offering this course to employees who do not understand basic computer concepts would more than likely result in a failed training intervention. This means retraining would be necessary and the organization would have to spend additional dollars for the effort. In this example, the manager's decision helped to avoid this scenario.

Creating an environment for success occurs when you look at the characteristics of your target population and identify the skills that learners must have before instruction begins. This can result in a successful instructional experience.

Assessing skill levels is crucial to the training development process. Successful organizations take the time to find out about their audience. In most instances, failure to perform this task results in a failed training implementation, possible retraining, and loss of training dollars. As a Trainer, before you teach a topic, always ask the question, *"What does my learner need to know first?"*

Understanding How Adults Learn

In addition to having varying prerequisite skill levels, learners come to training with a variety of characteristics, attitudes, perceptions, learning styles, and processing abilities.

What Are Learning Styles?

As stated previously, adults have different motives for learning. A learning style is a preferred way that a person learns most effectively and enjoys it. As a Trainer, it is important to have an awareness of learning styles. Adult learners may possess a few preferences for each style. It is important to use a variety of instructional methods (discussed in Chapter 5) to appeal to three of the main senses used in training: seeing, hearing, and feeling. There are numerous learning models and theories developed by experts such as David Kolb, Heil Fleming's VARK, Honey, and Mumford. There are also learning style inventories that allow a person to quickly identify his or her style preference. I like these three simple ones.

Visual learners like to see it! Visual learners will try to imagine or picture something. They prefer information presented visually or in a written format. Visual learners are often easily distracted by things going on around them. A visual learner prefers face-to-face contact rather than speaking over the phone. Visual learners prefer diagrams, demonstrations, maps, and pictures.

Auditory learners like to hear it! Auditory learners apply listening skills. The phonetic approach works best for this style of learning. Many auditory learners enjoy talking as much as they do listening. These learners are distracted by sounds and loud noises. Auditory learners prefer listening to lectures, audio tapes, and participating in group discussions.

Kinesthetic learners like to touch it! Kinesthetic learners enjoy learning by feel, touch, and hands-on activities. A kinesthetic learner prefers discovery learning and learns by being part of something. Those who never use step-by-step instructions, but instead jump right in to put together a newly-purchased computer table are most likely kinesthetic learners.

During your instruction, you may have to offer different types of practice exercises to accommodate learners with various learning styles. This approach to teaching emphasizes that individuals perceive and process information in very different ways. Understanding learning styles, profiling your audience, and identifying prerequisites all contribute to creating an environment for success.

Processing Information and Learner Recall

Some adult learning theorists believe that if you don't use new information within 48 hours, you will lose it. In addition to understanding motives for learning and learning styles, adults also process information differently:

- Delayed processing is when you ask a question and the person needs to take time to think about it—the response may not be immediate.

- Bilingual language processing or dual processing occurs when a person must hear the words in the conversational language used by the Trainer, translate it to their native language, and then back to the conversational language in order to respond—the response may not be immediate.

- Right versus left brain processing describes how the two hemispheres of the brain process information differently. Right brain individuals are said to be more intuitive, random, and see the big picture. They represent your creative people such as artists, writers, and musicians who process the world differently. Left brain people, on the other hand, are thought to be procedural, structured, methodical, orderly, logical, and analytic, and see only parts (not always the big picture).

What Issues and Barriers Can Affect Learning?

Nicole has worked in the Accounting Department for 10 years. She is a valued and trusted employee. She has been selected as one of five "Train-the-Trainer" participants. She will become a super user after the system is implemented. Nicole is excited and eager to learn the new financial management system. With her pen and paper in hand, she attends the first class which is "Creating a Purchase Order." She does well with the first two topics. However, when the math section begins, she feels a bit intimidated. She must learn how to manually calculate the units, quantities, rates, totals, and make adjustments. Nicole panics because she was a poor math student and never mastered it. She immediately goes to the bottom of her brain (the place where most humans go when stress or trauma occurs—the frontal lobe shuts down). She becomes fearful of her performance and she is embarrassed because her supervisor is in the room. Will this barrier have an impact on her success?

As a Trainer, when adult learners come into your classroom, unless you have completed a detailed audience analysis (discussed earlier), you know nothing about them. You know nothing about their cognitive abilities, emotional issues, or stages of development. In addition, you will know nothing about their past educational experiences. For example, have your adult learners:

- Endured overcrowded classrooms

- Come from disadvantaged communities where educational opportunities were sub-standard

- Practiced good study habits

- Exhibited deficiencies in basic skills

- Acquired good listening skills

- Learned about gender, race, and diversity issues?

Students can encounter barriers that can have an impact on learning. Corporate universities were started to support change, retain employees, and to provide professional development. Courses in basic skills (reading, writing, and mathematics) are offered in the workplace by many corporate universities. As a Trainer, you can't always pinpoint where learning issues originate in adults. Unfortunately, you will not have the time or expertise to fix some of the problems. However, as a Trainer, there are a few things that you can do (discussed in Chapter 4) to help identify learner issues such as body language, non-verbal behaviors, and responses to questions.

Responding to Barriers

There are countless barriers that can keep adults from learning. These include child care issues, deadline of time pressures, transportation challenges, financial problems, employment obligations, lack of self confidence, or and intimidation, to name a few. Here a few suggestions:

Some learners believe they can no longer be productive because of their age. They might be anxious about learning new skills, adjusting to rapid technological changes, and keeping up with workplace changes.

Age is not a barrier to learning. Allow adequate time for development of the new learning. Reinforce the value of learning the new skill. Actively involve the learner in the learning process so it can become meaningful. Self-confidence reinforcement, praise, and feedback must be given continuously.

Consider the past educational experiences of the learner. For example, has there been a previous lack of interest in learning? Is there a lack of confidence in the learner's own abilities?

If past educational experiences were negative, a variety of instructional methods (discussed in chapter 5) should be used in conjunction with self-paced independent learning, discovery learning, computer-based training, or one-on-one instruction might alleviate fears and anxieties.

It may take longer to accomplish a task for fear that it may not be performed correctly.

Allow more training time to learn new material and include adequate practice time.

Some learners might have slower reaction times when learning new tasks, a decline in manual dexterity, vision problems, issues with short-term memory retention, or recall difficulties.

Use repetitive strategies such as practice and review. Positive reinforcement and praising good behavior helps to build self-confidence. Be sure to include longer periods of study, practice time, and clearly defined strategies for retention, recall, and application of the information.

Learners process information at different rates.

Be patient. Check for understanding by using a variety of questioning techniques (discussed in chapter 6). Watch the non-verbal behaviors to determine if a transfer of learning has occurred.

Managing Diversity in the Classroom

Before you can make your training events memorable for your learners, you must manage your own preconceived notions, prejudices, and beliefs. The job of a Trainer is to create a non-threatening safe environment where people feel welcomed so positive learning outcomes can be achieved.

Let's review an end-of-course evaluation from a student.

> The Trainer would not look at me! I'm a Vietnamese American born in the U.S. I do not have an accent. I work in Marketing and Sales. Because I was a top earner in the company, I was sent to a management training program to help new managers learn people skills. I spoke to the Trainer several times via phone to gather the pre-class materials so I would be ready the first day. He was helpful and I was looking forward to his class. We corresponded via email, had a few conference calls with other participants, and two e-meetings. We even talked about our kids and like the same movies. We made plans to go to the theater once we got to Chicago because it was our first visit to the city. When I met him in person, his facial expression was of shock and disbelief. During class, he wouldn't even look at me—even when I raised my hand to respond to questions. During lunch break, I asked him about his obvious negative behavior towards me. He told me that in 1967, his grandfather was a prisoner of war (POW) in Vietnam and that he lost many of his friends during this era. He said he had been taught at an early age not to trust or interact with anyone from my native region. I asked him if he would be open-minded and willing to get to know me better, but I received no response. I believe that this Trainer should not be in the classroom because his hatred for the Vietnamese is so deeply ingrained that he cannot control it.

When attitudes and opinions are deeply ingrained, *stereotype rigidity* is the result. When clear evidence has been provided that the stereotype is false, but the individual is still resistant to changing his or her opinion, the worst form of prejudicial treatment is often apparent.

Trainers are human beings, too! At some time, everyone has made assumptions and stereotyped people based on preconceived notions. It could be someone's hair style, gender, race, clothing, body image, or age. However, in your quest to become an effective Trainer, you must challenge your perceptions and avoid negative filters that can affect how you perform in the classroom. Your goal is to:

* Examine your assumptions and stereotypes

* Be open with yourself and others

* Learn something useful about each learner that you encounter.

Why Is Understanding Diversity Important?

A truly diverse classroom or workplace includes a mosaic of people who bring a variety of backgrounds, styles, perspectives, values, and beliefs into the environment. On a day-to-day basis, we incorporate a variety of differences in the workplace such as educational levels, learning styles, problem-solving abilities, communication styles, job skills, talents, and personality traits.

How are we the same?

As a Trainer, keep in mind that we all have a lot in common. For example, everyone has the following needs for their families and themselves:

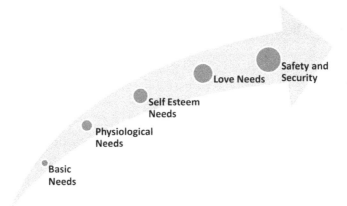

How are we different?

Diversity refers to the differences among people such as:

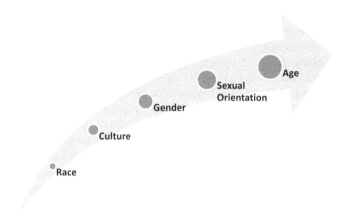

Race, gender, and ethnicity (customs, languages, common history, and nation of origin) are the visibly recognized forms of diversity. There are other forms such as people with disabilities, non-traditional family structures, ethnic cultures, sexual orientations, social classes, religions, and occupations. Trainers must recognize the differences, how they can influence the classroom, and how to manage training with a diverse group.

Inclusion in the Classroom and Its Value

No one wants to feel left out, targeted, or isolated! Inclusion means welcoming everyone's talents, skills, and competence to do a job regardless of race, age, gender, sexual orientation, religion, ethnicity, or physical ability. Promoting diversity and ensuring inclusion are critical factors for a successful classroom. As a Trainer, your job is to teach well and it's crucial to teach without bias. Part of creating a non-threatening, safe environment is to include everyone. This means making eye contact, interacting, establishing positive body language, fostering involved instruction, responding to questions, and making every individual feel part of the group.

As an effective Trainer, inclusion allows learners to stretch their skills and capabilities. For a learner, inclusion establishes a sense of belonging, feeling respected and valued for who you are, and feeling supported by committed co-workers.

Dealing With Your Perceptions

Perceptions are formed and enter our lives through parenting, community, experiences, and backgrounds. Perceptions are also formed by what we see on TV, what we read in the newspaper and online, what we hear on the radio, and the advertisements that bombard us.

The origin of prejudice, subtle discrimination, and the relationship among prejudiced attitudes and behaviors often come from how our perceptions were formed. As a Trainer, your job is to create an atmosphere that is conducive to learning. Take an active stand and identify when you:

- Prejudge an individual or group

- Hold strong ideas about certain people, their culture, or their religion

- Have been influenced by family, friends, and the media to formulate ideas about something or someone with whom you have no personal experience.

As a Trainer, your perceptions must always be challenged.

Are You a Diversity-mature Trainer?

A diversity-mature Trainer realizes that people have different backgrounds, perspectives, opinions, and objectives. Sometimes these differences can result in positive or negative interactions. The goal is to accept that differences exist—and then accept the people who have them. The diversity-mature individual challenges stereotypes, is open to differences, has a willingness to learn, and makes the mindset shift necessary for positive results.

How well do you rate yourself on diversity issues? Let's begin with a quick self-awareness questionnaire.

Self-awareness Questionnaire	Never	Seldom	Always
1. Do you think about how your statement may be interpreted, before you speak?			
2. Does your training delivery say one thing about learners, but your body language something different?			
3. How often do you stereotype learners when they come into your classroom?			
4. Are you open to learning about backgrounds and cultures that are different?			
5. Do you challenge your own perceptions and avoid making assumptions without facts?			
6. Can you show respect for every human being? Do you always abide by the golden rule: do unto others, as you would want them to do unto you?			
7. Are you patient with your learners? Do you validate your observations, interpretations, and evaluations before you jump to conclusions?			

Self-awareness questionnaire

To score the self awareness questionnaire, if you checked "always" for questions, 2, and 3, you should re-evaluate your thinking as a Trainer. Likewise, if you checked "never" for questions 1, 4, 5, 6, and 7, it might be a good idea to re-evaluate training as a profession.

This next example is not related to adult-learners, but the impact is certainly worth discussing. This example describes how powerful you can be as a Trainer by creating a safe learning environment and using alternate methods of engagement.

In the movie *Freedom Writers*, the classroom was an integrated yet segregated group of 14-to-15 year-olds who were angry and resistant, in addition to being typical rebellious teenagers. When the students walk in, the teacher is amazed as she watched the dynamics of the group. Each ethnic group (Cambodians, White, African-Americans, and Hispanics) gravitated to and sat with each other. She decided to integrate the segregated classroom. She used an experiment by asking a question of the group. The students who responded "yes" to each question had to move forward on the line. This experiment let them see that even though their ethnicities were different, they were alike in many ways. This experiment, while piquing their interest, still did not win her students' trust.

In another scene, an unplanned training event actually created a memorable training moment. During class, the teacher scolded a few students for drawing a negatively charged picture. In the midst of yelling at them, she mentioned the word "holocaust" and someone asked, "What is that? From there, the learning dialog just opened up and began to blossom. She took them to the Holocaust museum and allowed them to meet a few survivors who shared their experiences.

She then decided to use the instructional methods of reading and journal writing to help her students deal with their day-to-day life experiences. She also assigned them to read the book *The Diary of Ann Frank*. She convinced them that because Ann Frank was able to write down her feelings through journaling, they should also keep a journal of their issues, challenges, triumphs, and emotions. This opened up another new dialog. Through this process, she was able to *permanently* win the respect and admiration of her students.

She created memorable training events that transformed, changed lives, and enlightened her students. By creating a safe, nurturing, and non-threatening environment coupled with her continued effort to show care and concern, she was able to close the segregation gap; interaction, friendships, and cultural understanding were the result.

When you effectively create an environment for success, understand how adults learn, consider the issues and barriers that can affect learning, and manage diversity in the classroom, you contribute to the success in creating memorable training events for your learners.

Now that you understand the role of instruction, the role of the learner, and your role as the Trainer, the next chapter focuses on how to use a four-step training model for instruction.

Using a Training Model for Delivery of Instruction

Before you can make your training events memorable for your learners, it's essential to use a training model for the instruction. As stated earlier in Chapter 1, there are three learning domains such as cognitive (mental skills or knowledge), affective (attitude, feelings, and emotions), and psychomotor (manual or physical skills). In most instances, it doesn't really matter what type of psychomotor skills you are training someone to do. You could be training someone how to use a computer program, how to bake a chocolate cake, how to ballet dance, how to plant vegetables, how to tie a shoe lace, how to play the piano, how to start a campfire, or how to operate brakes on a car—there is a process that should be used for the instruction.

Using the Four-step Training Model

I think of training as a pyramid because each topic builds on the previous one. There are so many models used in the training industry. The one that I prefer to use is a four-step process to deliver training to adult learners. *Please note*, this model (or variations of it) is currently used in the industry. I have modified it slightly and added my own features to make it functional.

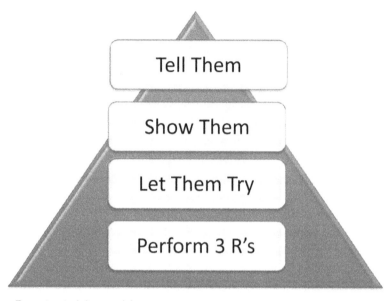

Four-step training model

Step 1 - Tell Them (Motivate) – Motivate learners by telling them what they will learn at the beginning of each chapter and topic.

Step 2 - Show Them (Instruct) – Instruct learners by defining, explaining, and demonstrating.

Step 3 - Let Them Try (Practice) – Practice, practice, practice! Reinforce what you taught them with practice and repetition.

Step 4 – Perform the 3 R's – Review, recap, and reward your learners.

Each step in this model should be used for each topic that you teach. For example, if there are six topics, this model will be repeated six times.

Let's look at the details of each step.

Step 1—Tell Them (Motivating Your Learners)

Why Are Learning Objectives Important?

As the Trainer, ask yourself: *"What does the learner need to know?" "What do I want to teach them?"* You can begin this process by telling learners what they can expect to learn by stating the *learning objective*. A learning objective identifies what the learner will be able to do at the end of each instructional unit. For example, *at the completion of this workshop, you will be able to create a purchase order using the Lynch Payroll system with 100% accuracy.*

A learning objective is observable and measurable and offers a great way to motivate the class. For the learner, it provides a clear picture of the course expectations. For the Trainer, it provides the opportunity to capture and engage the learner's interest in the concept being presented.

After the instruction, the learning objectives will be measured against learning outcomes. This can be accomplished by using a variety of instruments such as tests, questioning techniques, and demonstration. Whatever instrument you use, it must relate back to the learning objectives. If you haven't taught them, how can you test them? (Chapter 1 provides details about learning objectives.)

Why Explain the Benefit and Value?

In addition to telling learners what they will learn in the form of learning objectives, it's important to explain the benefit and value of learning the new information. Instruction can be easier to convey when adults are motivated by the benefit and value. Motivation is generated by incentives such as professional advancement, promotions, recognition by peers, an interest in learning something new, pay increases, or awards. As a Trainer, be sure to help motivate participants by explaining the benefits and values.

When learners recognize the usefulness of learning, this is a sure way to motivate them to achieve a positive learning outcome. The more they see the value, the more motivated they are to learn. For example, a value could be, *"With the new Lynch Payroll system, creating a purchase order will be easier, faster, and produce fewer errors."* Remember, Chapter 2 showed how learners are motivated if they see a benefit in learning the new information.

Step 2—Show Them (Instructing Your Learners)

As the Trainer, ask yourself: *"What is the easiest way to convey the message?"* and *"How can I make this clear for my learner?"*

Several years ago, my car was constantly getting stuck in the snow. Although I had no problems driving in the snow, it was always difficult for me to remove it from a parking space. On this particular day, my sister Lorraine was the passenger. After ten minutes of trying to maneuver my car out of the parking space, she asked, *"Do you know how to drive in the snow?"* I responded, *"Of course I do, I've lived in Chicago for 40 years."* She said, *"Show me."* After observing me, she said, *"Let me show you how to maneuver the wheels."* After I observed her instruction, I returned to the driver's seat to demonstrate what she just taught me. I was able to position my wheels correctly and remove my car without getting stuck.

In this step, begin this process by showing learners what they will learn through defining, explaining, demonstrating, drawing, and describing. Introduce the skills and concepts necessary to complete the task. Adults tend to respond positively to information if they see it, so demonstrate new information whenever possible. You can help participants understand the content by using clear instruction, applying presentation techniques that appeal to a variety of senses, and by encouraging interaction. Use the following steps to instruct your learners:

1. Define, describe, or explain the new material. If you are teaching concepts, provide adequate definitions. If necessary, illustrate or draw the procedure or process.

2. Demonstrate how to perform the procedure. For example, if you are providing computer training (or any type of psychomotor activity), demonstrate the task first as your learners watch you.

Then, allow learners to imitate you by following the steps on their own computers.

3. Use real life examples whenever possible. You can reinforce the learning process by using real life examples and scenarios with which learners are familiar.

Step 2 of the training model is when you TEACH and create memorable training moments! Let's take a quick look at technology training.

First, you should SHOW your learners by demonstrating the process while the participants watch you. For example, *"Let me enter a purchase order while you relax, watch, and take notes."* During the demonstration, walk through each step while providing explanations and definitions. Learners can visually see the process and make the connection of what is required. In most instances, a Liquid Crystal Display (LCD) projector is used in the classroom to display the system details while participants watch.

Beverly Rico, M.A., Training and Development, uses a unique approach to begin her training classes. Before participants log onto the system, she shows a simulation of the content using an interactive software program. Using captured screens of the new system, the mouse movements are demonstrated for a specific process. In her opinion, this method allows her to engage her learners in the beginning and it helps to minimize learners surfing the web, texting, and multi-tasking during the initial instruction (which typically results in interruptions and asking questions). After the simulation, then she allows learners to logon and watch her demonstrate the process via the projector.

Then, after your demonstration, learners will **imitate** you as you walk them through the process. They will follow along as you prompt them on what to do. For example:

*   Click the Unit field and enter . . .

*   Move to the Price field and enter . . .

*   Click the OK button

- The XYZ system responds with . . .

- Click the Close button

Remember, just because learners are imitating you does not mean they understand the content. Any Trainer can tell someone what to press, click, or enter. This is simply rote learning and doesn't always result in a knowledge transfer.

Next, as your learners imitate you (or practice on their own which is discussed next in Step 3 of the training model), it's your job to continuously check for understanding—I like to think of this as competency checks. You can ask questions such as: *"How is everyone doing?"* or *"Does this make sense?"* or *"Is everyone on step seven?"* During your instruction, avoid asking a direct question, because you don't want to embarrass someone or cause undue stress. Remember, adult learners come into the classroom with educational backgrounds, personal challenges, and day-to-day situations that are unknown. Be careful with questioning (discussed later in Chapter 6.)

Let's look at a few techniques used for computer training.

Techniques for Computer Training

> I was assigned to evaluate the training delivery of a few new Trainers. I don't believe in evaluating anyone unless they have taught the class at least three times. By then, the Trainer will be familiar with the content and chances are that most of the questions tend to be the same. On this occasion, the Trainer was demonstrating the task and explaining the concepts, while the learners were trying to keep up with him. While he was on step 11, two people were on step 7, three were on step 5, and only one person was on step 11. Every person had the training manual open, but he didn't refer back to it. He sat down the entire time while eating a donut, used his laser pointer, and had no clue if the learners understood the content. Learners were frustrated and disappointed.

During Step 2 of the training model, you will show learners by demonstrating the new system function. Computer training uses memorization, listening, and psychomotor skills to perform the system functions. Computer training can be a highly controlled activity because it involves the Trainer standing in front of the classroom telling the learner what to press, push, or choose. In the classroom, the learner:

- Imitates the demonstration provided by the Trainer

- Follows step-by-step procedures

- Performs blind repetition.

Avoiding Blind Repetition

When learners imitate the Trainer, this is simply known as blind repetition. For example, you tell the students, *"Click the insert menu"* and they do it. Then, you tell them to, *"Press the enter key"* and they do it. Next, you tell them to, *"Click the file menu,"* and so on. This step-by-step process proves to be effective in teaching computer applications. However, it also promotes dependence. Using this controlled method, each learner follows the Trainer, and the Trainer can monitor the class because everyone should be on the same step with the same results. However, you are simply walking them through the steps telling them what to push, press, or click while they perform the task. Memorization of steps does not translate into understanding the content or the logic of the application. There is no guarantee of comprehension unless learners are involved in the process.

A good Trainer can encourage independence by asking questions, providing assistance, and prompting for input. Instead of making learners follow procedural steps, allow them to use intuitive processing. For example, an open question to gather insight is: *"What menu should be used to delete a purchase order?* To provoke critical thinking skills you could ask: *"What does this screen say to you?"* or *"What data is required in the name and address fields?"* Challenge your learners to think and they will become involved.

Enhancing Instruction through Movement

As a Trainer, what type of learners do you want in your classroom?

As a Trainer, keep it moving! With computer training, the Trainer provides information and the learner is often a passive receiver. To avoid this traditional, authority-based approach, Trainers should walk around the room, build rapport, and look at what the learners are doing. Movement increases interest level. It also can allow you to help someone without embarrassing him or her. A technology Trainer should NEVER, EVER sit down while training. It is important to interact, engage, and identify possible learning gaps during this time. I like to position myself in an area where I can see everyone. Whenever the client lets me, I set up my classroom in a non-traditional way using a u-shaped layout (discussed later in Chapter 8). This way, I can zoom in on an individual and scan the room in a matter of seconds.

Using Questioning Techniques for Competency Checks

Questioning techniques (discussed in Chapter 6) help you assess knowledge transfer. You can quickly identify if the content is understood and if learning has actually occurred. Avoid asking your class, *"Does anyone have any questions?"* Instead, provoke critical thinking skills by asking a specific question. For example, *"What are the two ways to delete a purchase order?"* or *"How can the purchase order be adjusted in the XYZ screen?"* Training is a controlled event so learners don't always "know what they don't know." As a Trainer, you must ask clear and concise questions.

Questioning techniques such as closed, open, leading, next-step, and reflexive can be powerful tools when teaching computer novices because various types of questions can be used to identify whether or not learning actually occurred.

Using Discovery Learning

The discovery theory states that a learner, through his or her own effort, seeks out new information. In some instances, the content will be remembered better and longer than if a Trainer simply provides it through a more traditional means such as lecture or demonstration.

A good Trainer uses methods that allow for discovery learning. For example, you could say, *"I know this is the first time viewing the purchase order screen, but what do the fields tell you?"* The most valuable and lasting learning is when the learner discovers the answer. This is when a learner can achieve an "Ahaaaaaa" moment.

As a Trainer, empower your learners to learn on their own. Don't just tell them and show them; challenge your learners to think. A learner remembers more if he or she figures it out. Allow learners to take responsibility for their learning. During discovery, a learner subconsciously believes the information is important.

Using Comparison Learning

Learning is work! Learners will not invest any energy into learning something new unless they understand the benefit. It is a good idea to present new information by connecting it with something the learner probably already knows. A comparison can relate the new topic to a known topic such as:

- In the old system, you entered a purchase order record as . . .

- *In the new system, you will enter a purchase order as . . .*

- In the old system, it took 10 steps to enter a purchase order.

- *In the new system, it will take 5 steps to enter a purchase order.*

Comparisons are an effective way for learners to comprehend the value of new information.

Teaching from Simple to Complex

Never start with the challenging material unless your goal is to lose or intimidate your learners. Remember, learners are "innocent" users who (in some instances) have been forced to learn new information. Your job is to create a safe environment to minimize learner frustration. When teaching a new topic, build on what learners already know. To succeed, teach simple concepts first, then move to more complex topics. For example:

- If you are teaching a word processing course, teaching learners how to create a letter from scratch is easier than using a template (a template requires prior knowledge).

- If you are teaching an email course, teaching learners how to compose an email is easier than sending an attachment (an attachment requires prior knowledge).

Be sure to teach simple concepts first to allow learners to build on their knowledge. When topics are complex, repetition is a powerful learning tool. If your learners see, hear, and experience something repeatedly, they will remember it better and longer.

Using Precise Terminology

I attended a class where the Trainer would make statements such as, *"Click the button on the bar at the bottom of your computer."* Or, *"Click the box at the top of the window."* Be sure to use exact terms and the right terminology such as icon, button, menu bar, title bar, scroll bar, function keys, tool bar, task bar, radio button, and link. Never tell learners incorrect information. For example, what the Trainer should have said was, *"Click the start button, on the left side of the gray task bar, located at the bottom of your desktop."* Or, *"Click the maximize (middle) button located on the right side of the title bar.*

Invest the time to learn the correct terminology; it will reinforce the learner's knowledge of these terms for future training, it will give you more credibility in the classroom, and *it's the right way to train.*

Never Do It for Them!

I n my early days of computer training, I was teaching a group of employees how to use a word processing program. A gentleman asked me a question and I took the mouse from him and did it for him. He said, *"I'm not stupid, I'm a novice. I have the ability to learn, and I don't need you to do it for me—just show me."* Boy, did I get a lesson that day. I surely didn't mean to offend him, but I did. Of course I felt bad and apologized profusely after class. In Step 3 of the training model (Let Them Try) is when you allow practice while you monitor and provide help.

Never touch a learner's mouse or keyboard. Often during computer training, you will come across learners who are having difficulty performing a task, completing a class exercise, or are slow in learning to use the mouse. However, a Trainer should never grab the mouse and take over the task. Use this as an opportunity for a discovery learning experience. Be patient and assist the learner through the process. It may take some time to fix the problem, but don't rob the learner of a discovery moment.

Take over only when it is absolutely necessary (the system shuts down, slows down, keyboard is frozen, or it's not obvious what happened). Even if this occurs, still allow the learner time to troubleshoot the problem first. You can verbally tell them what to do and watch them try to solve the problem. If this does not work, you can intervene. This also applies to fellow participants who may offer help to their neighbors. It's your job to assist the learner. NEVER DO IT FOR THEM! Remember, we are all novices at something. Don't confuse being a novice with incompetence.

Step 3—Let Them Try (Practicing New Skills)

As the Trainer, ask yourself: *"Did the light bulb go off?"* or *"Was an 'Ahaaaaa' moment achieved by the learner?"*

In Step 3, allow learners time to practice the new skills you just taught them. Learners will work independently while you watch and monitor their performance. Practice provides self-directed as well as discovery learning. Be sure to walk around the room to monitor progress, answer questions, develop one-on-one relationships, and provide immediate feedback. As you walk around, you can identify who is having problems and how you can adjust your training to help them.

With computer training, be sure to use a simulated "test database" that provides real "live" client data. The test database should include case-based scenarios and real life examples. During practice, your learners can use hands-on practice with real simulated data that they can relate to. For example, if you are teaching learners to create purchase orders in a manufacturing environment, don't use data from a retail environment. Your test database should always contain relevant data with which learners are familiar. It's important to relate new learning to something learners already know (discussed earlier in Chapter 2).

Use the following steps to help your participants practice new skills:

1. Provide practice, repetition, and recalling techniques.

2. Answer questions and provide feedback.

3. Pay attention to learners! Watch what they are doing to determine what they need.

During this step of the training model, learners will work on their own to simulate realistic situations. As a Trainer, you can provide feedback and reinforce information. Tell learners when a task is done correctly and help them understand when something is done incorrectly and why.

One of the adult learning principles discussed in Chapter 2 is that short-term memory decreases as people age. Therefore, repetition and practice help adult learners retain new information. Participants can reinforce new information by practicing the new skills they have learned. Adults need repetition and techniques that help them recall information.

Step 4—Perform the 3 R's (Reinforcing learner performance)

As the Trainer, ask yourself: *"Can my learners return to their jobs with the ability to perform the new system functions learned during my training class?"*

The final step of the training model is to *review*, *recap*, and *reward* what was taught. The Trainer summarizes the new information and helps participants understand how they can apply their new skills to real-life situations. You can help identify future applications by asking learners to suggest how they intend to use their new skills.

Use the following steps to perform the 3 R's:

1. **R**eview the lesson's content and details. For example, today you learned how to enter a purchase order. Ask for and answer questions. Use discovery learning techniques. For example, you could ask the class an open question such as: *"What is the first step to search for a purchase order that was created six months ago?"*

During the review section, perform competency checks to ensure learners understand the content. Use questioning techniques (discussed in Chapter 6), classroom assignments, demonstrating a task (hands-on exercises), resolving case scenarios, or troubleshooting issues. During practice and review, determine if an increase in knowledge or skill has been achieved.

2. **R**ecap the lesson. Summarize the main points and emphasize any areas that may be assessed.

3. **R**eward your students; make them feel good about the learning experience.

This step in the model brings a sense of closure to a lesson. The four-step training model should be used for each unit of instruction (chapter and topic) you are teaching. Present the material as clearly as possible and at an appropriate pace. Be sure to schedule time for learners to practice. As you complete each phase of training, remember to review what was taught and reiterate the benefits of the new learning.

Let's review an end-of-course evaluation from a student.

> I left this class so confused. The course was supposed to teach us how to use our new email system. I know what I came to learn, but I don't feel like I learned it. I followed the agenda, but the Trainer never explained what we were doing or why we were doing it. He stood in front of class and told us to follow along. The instruction was extremely confusing. The Trainer never stopped and asked if we had questions or needed clarification. I was excited about the topic, but I left feeling annoyed because I wasted a precious day. The flow was not logical and we never practiced anything. I will definitely have to use my valuable time to take this class again. My end-of-class evaluation will not be rated favorably.

Using the four-step training model for your delivery allows you to provide quality instruction. The next chapter will focus on how to use effective presentation skills for your training delivery.

Presenting Topics and Delivering Training

Before you can make your training events memorable for your learners, evaluating and improving your delivery are key. Your delivery can only be effective if you are well prepared and present the topics in a clear and logical fashion—but don't ignore the human factor.

When learners meet you, they should feel without the words being said, *"I care about you!"* It is important to display an interest in your learners' development. You can have glossy laminated beautiful handouts, expensive donuts and coffee, and a lavish training room with state-of-the art equipment. However, none of these items will compensate for inadequate preparation or ineffective presentation techniques.

Let's look at a few examples of how skills and behaviors were changed as a result of an effective training style. The examples on the next page do not specifically relate to adult-learners. However, the point is that successful learning outcomes occurred because of effective training delivery. A Trainer can improve the quality of his or her delivery by identifying learners' hidden needs, using a variety of instructional methods to meet varying learning styles, tapping into their unknown abilities and interests, and using resources to assist in instruction. Whatever it takes to engage your students and create memorable training moments—use it!

It helps when learners are excited too, but often the teacher has to stimulate the desire to learn. In the movie *Take the Lead*, the teacher used a simulated dance move to entice the bored, unmotivated students. He showed the benefit of how a different genre of dance could expose them to a different world. In the movie *Mr. Holland's Opus*, the teacher used popular music to win over his uninterested students. In the movie *Lean on Me*, by showing care and concern for his students' well being, the principal was able to inspire them to achieve. In the movie *Stand and Deliver*, the teacher aroused his students with AP Calculus. He tapped into hidden abilities, taught them that failure was not an option, and used pride to stimulate his learners. In the movie *Sister Act 2*, Sister Mary Clarence tapped into the creative musical interest of her students and watched them blossom into great singers. By using effective training delivery, each of these fictional teachers figured out a way to reach their learners.

Trainers have challenging jobs because they have to be motivating, stimulating, and interesting the entire day. This can make for a long and tiring day. In addition, if you determine that certain participants don't have the prerequisite skills and can't keep up, you have to work harder because you may need to give them individualized one-on-one instruction. I've worked with gifted Trainers who will stay late, give up lunch breaks, and even go to a learner's job (on their day off) to help them. As a Trainer, even though you may have a long and tiring day, keep in mind that when learners arrive in your classroom until the time that they leave, you are the role model, SME, and the person from whom they are seeking information.

Learners expect a Trainer to demonstrate expertise. An audience responds favorably if the Trainer appears to enjoy the art of training, seems trustworthy, and shows care and concern. You can encourage interaction and present topics effectively by using natural gestures (smiling and laughing) and body language, injecting humor when appropriate, using real life examples and analogies they can relate to, creating visual aids to illustrate main points, and providing a variety of instructional methods (discussed later in this chapter) to stimulate learning.

Demonstrating Effective Presentation Skills

Speaking effectively is vitally important for your training delivery. Your entire body should be used when you speak publicly, not just your voice. You must consider making eye contact, your body posture, and speaking rate for your training delivery.

Managing Eye Contact

As you speak, face your learners so they can hear you. A soft focus on the entire face, not just the eyes, is sufficient. Scan the room consistently and try to make eye contact with everyone. Try to avoid being fixated on one person. Maintaining good eye contact means looking at your learners (not staring) while you move around the room. As you make eye contact, be aware that your facial expression should be warm, friendly, and approachable. Be sure to smile and nod at appropriate times to let people know that you are listening, attentive, and care about them. Never look away, look around you, or do something else when a person is speaking to you. This lack of eye contact may make you appear rude and suggest that you'd rather be somewhere else.

Many people feel that the eyes provide a revealing way to communicate. For example, what do these facial expressions tell you?

- Positive and attentive

- Makes no eye contact

- Tired eyes

- Narrowed and shifty eyes

- Eyes move slowly

- Rapid eye movement; too much blinking

- Looks away while you talk

- Relaxed and confident

Establishing Body Posture

It's important to project your voice in a clear manner. Typically, you can speak from your diaphragm if you stand in the *"at-attention"* military stance. You can project your voice better if you speak from your diaphragm. This is very important because after a long period of training (five or more days) you can achieve fatigue in your vocal cords. Be sure to use the proper body stance and avoid speaking from your upper chest or throat.

It's important to interact and encourage participation. You can approach your learners, move around during lecture or in-class exercises, scan the group, and always face your learners so you can see them and they can see you. Never sit down while teaching.

Speaking Rate and Vocal Inflection

You can develop rapport by approaching your audience as you speak. Learners can easily become distracted so be sure to speak at a rate (not too slowly and not too quickly) that keeps participants interested in what you have to say. Try to avoid using pausal expressions such as "uh" or "ummm" as well as excessive repetition of words.

Your vocal inflection and volume offer another way to deliver quality presentations. When you are energetic and enthusiastic about a topic, it translates through your voice. Your audience is more likely to perceive you as knowledgeable when you speak vigorously. Use volume and inflection to draw the audience attention to important points.

Avoiding Inappropriate Mannerisms

There is an old saying—it's not what you say, it's how you say it. There are certain attitudes and mannerisms that must be avoided when interacting with your learners. It's difficult to be perfect, but try your best to always have the learners' interests at heart, and never appear preoccupied with something other than the topic. Other mannerisms to consider:

- Although you are the SME, you must avoid exhibiting egotism, vanity, or talking down to your learners. This may be challenging when dealing with difficult participants, especially those who challenge you. Training is an exchange and a collaboration, so avoid a closed-minded attitude or appearing overly persuasive about a subject. Remember, this is also an opportunity for you to learn, too.

• Movement builds rapport (discussed earlier in Chapter 4), so never hide behind your desk or podium, sit down while teaching, or remain in the same place too long.

• Be prepared and always have your handouts or instructor notes available. Avoid shuffling papers and looking for information.

• Always use simple plain language when teaching a technical topic. Remember your learners are "innocent" users.

• Avoid using acronyms, jargon, and appearing overly technical or analytical.

Communicating through Body Language

"The unconscious of one human being can react upon that of another without passing through the consciousness." – Sigmund Freud

While sitting at the computer, Mya has her hand to her cheek—this could indicate that she is thinking and evaluating what to do next. Damon's arms are folded—could this mean that he is angry or could he simply be cold? Deciphering someone's body language can be difficult because it might have multiple meanings. Even if the person does not speak, he or she is still communicating.

Body language is the non-verbal way that people's movements and gestures can convey how they are feeling and can sometimes indicate what they are thinking. Body language includes:

• Eye contact

• Facial expressions

• Body posture

• Hand gestures

• Tone of voice (inflection, volume, and pacing).

Ninety percent of communication is non-verbal. It can reveal what a person is feeling before actual words are spoken. As a Trainer, how you respond non-verbally is just as important as the words that you speak. A smile, hand-gesture, frown, look of confusion or anger, all "say" something about your true feelings. Body language is sometimes clear and unambiguous.

What Does the Trainer's Body Language Say to Your Learner?

Make sure your facial expression and body language match your message. For example, if your learners walk in the room and you are standing behind the podium frowning with clenched fists, how will this be interpreted? To create a non-threatening environment that is inviting, your body language must be warm and friendly.

In terms of percentages, the Trainer's body language, which includes facial expressions, eye contact, showing concern, body movement, and gestures represent what learners react to the most. Next is your vocal inflection—the changes in your tone of voice, volume, and the sound of empathy and understanding you communicate using your voice. Finally, the actual words that are spoken represent the lowest percentage. Your learning audience reacts more to how you present the material than to what you say.

Stimulate the learners' interest and illustrate what you want to say by using natural gestures with your forearms and hands. When you point to a visual aid, use your opened hand, not your finger. Avoid aggressive pointing or body posture, which can be misinterpreted. Avoid crossing your arms; instead, appear open or accessible to your learners.

What Does the Learner's Body Language Say to the Trainer?

Let's look at a scenario on reading body language.

During the practice session, you notice that Thomas is visibly struggling with the practice exercise. However, during the instruction, he answered questions, participated in the group discussion, and appeared alert, interested, and excited about the class. Thomas's body language is in conflict with your perceptions.

Body language and non-verbal behaviors can work together. Non-verbal behaviors include a range of unconscious physical movements. These behaviors portray a person's true feelings and can indicate how they are coping with a situation. By paying attention to a learner's non-verbal behaviors, you can learn quite a bit about their feelings.

Analyzing Your Learners During Instruction

As a Trainer, you must learn how to identify the reaction of your audience and observe their non-verbal behaviors. A skilled chef can intuitively identify a missing ingredient, a skilled personal trainer can instinctively gauge whether a person can perform another repetition in a set, and a seasoned and involved parent automatically knows if a child is not telling the truth. Likewise, a skilled Trainer can intuitively "feel" a question coming simply by the expression on someone's face. This can often occur even before the learner has formulated the question. As you become more efficient in the craft of training, you will be able to quickly identify confusion and/or excitement in learners by analyzing their body language.

Be sure to watch for non-verbal behaviors during each step of the training model. During Steps 1 and 2 of the training model, watch to see if learners' facial expressions exhibit understanding or confusion. During Steps 3 and 4 of the training model when they practice and recall the lesson, you can monitor their reactions. For example, during practice exercises a few learners are frowning while others are looking dazed and confused. What does this say to you? Ideally, you want body language that exhibits a high level of content expertise, self-confidence, and full participation.

As a Trainer, you must constantly observe your learners' reactions throughout the day to determine how they are responding to the training. When performing competency checks, you can identify audience reactions by listening to their responses and observing body language. A learner's body language might suggest a response, but your assumption may not always be accurate. For example, Nikki has completed the in-class exercises with 100% accuracy. However, when you ask her a direct question she gives you a blank look and offers an incorrect answer. Body language may not always match the given situation—a frowning participant with folded arms might signal the participant is frustrated or cold—but it does offer a clue about the learner's understanding of the subject matter.

When people don't put feelings into words or are unable to find the right phrases to describe their emotions, their non-verbal communication might indicate their true feelings. As a Trainer, you must tune into the body language of your learners. They will provide helpful hints about their progress and what is needed for their success.

Using Instructional Methods to Invigorate Your Training

Instructional methods are primarily used during Step 2 (Show Them) of the training model. During your training delivery, use a variety of instructional methods to appeal to a wide range of learning styles and processing abilities. Be sure to use a combination of techniques in succession to produce the most beneficial results.

Now that you know how to use the Four-step Training Model, let's look at the instructional methods that can be used during your instruction. The methods may be determined by the size of the audience, participant skill level, and the subject matter you are teaching.

Let's look at an example.

I attended an Emergency Preparedness Training Summit. The training event was for health care professionals, faith-based organizations, and first responders (police, fire, and EMS). The goal was to prepare these individuals on how to respond when a disaster (such as the H1N1 "swine flu" outbreak, 911, or Hurricane Katrina), mass casualty, or a pandemic occurs. The workshop dealt with the physical and psychological factors that can occur when people experience an emergency situation.

The workshop opened with a visual aid, a film that simulated the event. This powerful instructional method depicted the entire event of United Airlines Flight 232 that crash-landed in Sioux City, Iowa in July 1989. The Trainer was the pilot, Captain Alfred Haynes.

The Trainer provided an engaging and powerful training session by choosing film as the instructional method. While describing the event, he walked the learners through the entire 44 minutes of the event. As each episode unfolded, learners could visually see each incident. From the *inception*—when the pilots learned that an engine destroyed all three hydraulic systems, through each communication with the air traffic controllers; to the *completion*—when pieces of the plane broke away resulting in the emergency crash landing of the aircraft.

In this training session, a variety of instructional methods were used. Even though film was used as a visual aid, the Trainer also incorporated simulation, lecture, question and answer, and experience sharing. Using a variety of instructional methods was a great way to convey the training message and it helped learners quickly understand the content. Whatever it takes to engage your students and create memorable training moments—use it!

Using a variety of instructional methods allow you to create exciting and engaging opportunities in the classroom. Let's look at a few.

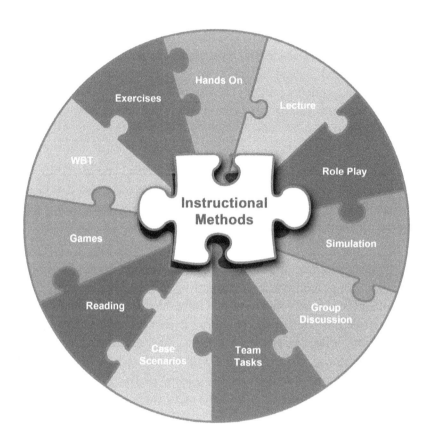

The table below provides a brief explanation.

Method	Description
Lecture	An oral presentation intended to teach a particular subject. Usually, the Trainer will stand at the front of the room and recite information relevant to the content. Learners are using listening skills and taking notes.
Demonstration and Hands-on	The Trainer demonstrates the task to the learner. Learners can visually see the interaction with the object. This is a useful tool when teaching psychomotor skills such as computer training.
Visual aids	Visual aids offer a great way to pique the interest of your audience. Pictures, films, movies, and videos allow you to clarify important points and multiply the understanding of the content. In addition, colorful handouts, job aids, and illustrations with callouts are a great way to support the content being learned. Words plus images effectively reinforce the material and allow learners to see, hear, and feel the content being conveyed.
Role play	Participants act out new roles learned. This interactive and collaborative activity helps to engage the imaginations of the learners.
Group discussion	The Trainer leads the group in a discussion about the new learning. Participants can share, collaborate, and learn from each other.

Instructional methods descriptions

Method	Description
Team tasks	Participants are split into teams to complete a task as it relates to the new learning. Learners interact, exposing themselves to the skills and views of each individual in the group. It encourages participants to experience working in a team setting.
Simulation	The imitation of something real, modeling of the actual event being learned is a form of experiential learning. Simulations are excellent choices for modeling a real-life situation.
What if analysis	The Trainer leads the group in a type of scenario that involves analyzing what the results could be. Learners can explore a variety of assumptions to solve a problem. This method encourages the use of critical thinking skills and thinking "outside the box."
Case study (Case-based scenarios)	The Trainer provides details about a real-life situation and participants are asked to make a decision or solve a problem concerning the new learning.
Panel discussion	Experiences or opinions are discussed with expert participants.
Reading	Participants use written materials or instructions about the new learning. In some instances, Trainers may have participants read during the class.

Instructional methods descriptions (continued)

Method	Description
Experience sharing	Participants share individual experiences regarding a topic to provide a better understanding about the content. Participants become resources for each other.
Question and answer	The Trainer directs questions to participants to stimulate discussion about a topic. Questioning techniques (open, closed, direct, or reflective) provide ways for the Trainer to determine if the content is understood.
Games	Educational games provide a creative and fun way to explore the new learning. By using games, learners can use a variety of senses to understand the content.
In-class exercises	Exercises are the most effective way for participants to practice what they have learned. This method aids retention through interactive involvement. Trainers can observe if a transfer of learning has occurred or if further assistance is needed.
Computer-based training or web-based training	Participants use this self-directed, independent method of learning with variable instructional paths. Learners often work at their own pace and set their own schedules. Depending on the design of the system, regular feedback on performance is provided at the end of each instructional unit and managed through a Learning Management System (LMS).

Instructional methods descriptions (continued)

Other Effective Presentation Techniques

Incorporating Humor

Be careful! What may be funny to you could be offensive to someone else. Keep in mind that cultural differences do exist in the classroom. Using well-timed and relevant humor can be effective. For example, you can use an icebreaker at the beginning of the class or when you notice learners are winding down. You can also tell a joke, display a cartoon or comic, or share a personal experience relevant to the topic. Be sure to use it in good taste and appropriately. Avoid using profanity, stereotypes, or discussing personal issues. I attended a class once where the Trainer was always talking about her husband and their personal issues. It is inappropriate to discuss personal matters in the classroom. Always maintain a professional image.

Using Visual Aids

Since most of adults are visual learners, you can stimulate learning by using visual aids. Visual aids such as white boards, interactive smart boards, training videos, and projectors that show presentations and images are excellent ways to communicate with your learners. Be sure visual aids are simple, easy to understand, and visible to all participants. If you use a flip chart or white board, write quickly, legibly, and use large enough letters so your comments can be seen in the back of the room. When speaking, face your audience as much as possible and avoid blocking or talking to the visual aids.

A well prepared lesson plan coupled with excellent presentation and delivery skills will help your learners' development. Inadequate preparation or ineffective presentation techniques can cripple your learning outcomes. In the next chapter, you will learn how to encourage participation through questioning.

ENCOURAGING PARTICIPATION THROUGH QUESTIONING

Before you can make your training events memorable for your learners, it's important to interact with them using a variety of questioning techniques. How do you check for understanding? Of course, the learner's non-verbals and body language offer the first clues, but questioning techniques provide the best method to identify understanding of the content. Although there are a multitude of instruments that can be used to assess your learners such as pre-and post-skill discoveries, classroom assignments, demonstration of tasks (hands-on exercises), or problem-solving cases, questioning provides the quickest technique. The Kirkpatrick Evaluation Model[8] (discussed in Chapter 8), uses four levels to assess and evaluate learning outcomes as well as the worth and merit of the training initiative.

---

8   Kirkpatrick Donald, *Another Look at Evaluating Training Programs*, American Society for Training & Development, 1998

Let's take a look at what happened at the end of a training day.

I was relaxing in the Trainers' lounge after a full day of training. There were eight other Trainers that were part of a 12-course rollout. Two of the Trainers talked about how slow the learners were: one called them stupid and the other said they asked too many questions. A few negative statements were made about their cognitive abilities. They laughed and joked. I was really offended and couldn't take it anymore. I said to them, *"As a Trainer, no question is stupid; it is our job to teach, not criticize, and to encourage participation. If the students knew the content, we wouldn't have a job. If a student asks a question multiple times, the content is not clear. As a Trainer, use it as an opportunity to measure your skills as an effective presenter. This may mean that you have to use a variety of instructional methods to convey the message. And, if you make the connection, you've done your job and created memorable training. If they ask a question, simply answer it."*

As discussed in Chapter 4: Using the Four-step Training Model, you will constantly perform competency checks throughout the learning event and the best way to accomplish this is by Q & A sessions using a variety of questioning techniques. These Q & A sessions should occur at the end of each topic, lesson, and chapter. Most importantly, they should be scheduled during the comprehensive review at the end of the course.

Encouraging Interaction through Questioning

Questions are the most effective way to interact with your students. By soliciting ideas and asking meaningful questions, you can engage and encourage participants to interact with each other as well. Those who actively participate are more likely to retain new skills. When you interact with your learners it shows you are interested and care about their progress.

During the Q & A phase, keep in mind that learners sometimes act differently if peers or supervisors are in the room. In some instances, you may even get a better response when people don't know each other. It's important to be mindful of how you ask questions, especially for those learners who have had negative learning experiences, lack confidence in their abilities, or have not been in a classroom in a long time. Being placed in a learning mode can be stressful for some learners.

Developing Various Types of Questions

After initial terminology and concepts have been introduced and understood, it is now time for questioning. Although you use questioning in Steps 1, 2, and 3 of the training model, Step 4 is a perfect time to assess your learner's understanding of the topic. This is when you will review, recap, and summarize the material. Be sure to have a list of relevant questions prepared prior to class.

Competency checks must occur throughout the learning event through questioning. This includes questions about the content or the learner's performance. For example, *"Tyler, how are you doing with lesson two?* Questions are also a good tool for measuring understanding. They provide a great opportunity for you to identify how your learners are doing and to give them immediate feedback. In Chapter 4, we talked about how blind repetition is simply the memorization of steps, but doesn't measure true understanding. If you want your learners to retain the information and recall it later, they must be actively involved. Questions encourage participation and collaboration with the group.

When it is time to start the Q & A session, **never** begin with asking the class, *"Does anyone have any questions?"* Instead, you ask the question! For example, *"What are the two ways to save a file?"* As a Trainer, asking a non-specific question such as *"Does anyone have a question?"* is not the best approach. Learners don't always "know" what they don't know. Therefore, you must ask the question to allow participants to use their thinking and problem-solving skills. Let's look at the types of questions that can be used in the classroom and a few examples on how to use them.

Closed question—Requires a yes or no response or one specific answer, such as:

- What keystrokes are used to change the font?

- Does your purchase order show the price computed?

Open question—Tends to require discussion and typically do not have a short answer, such as:

- Can you describe the three ways to search for a purchase order?

- Why is it important to approve purchase orders over $10,000?

Leading question—Offer a discovery technique because they guide the learner in the right direction, such as:

- When calculating the price of a purchase order, you can . . .

- In addition to searching for a purchase order by number, you can also search by . . .

Next-step question—Allow learners to do intuitive processing and stimulate thinking, such as:

- When you click the navigation menu, what is the next step to access the purchase order?

- After you view a purchase order, what is the next step to print an image?

Reflexive question—Supply a discovery technique and allow the learner to use critical thinking skills. The Trainer returns the question to the learner to encourage him or her to discover the answer on their own through recalling. For example,

- Learner asks: *"How do I access the purchase order system?"*

- Trainer responds: *"Let's return to the desktop and see if you can locate the icon."*

Using Questioning Techniques to Ask the Question

A Trainer can respond to learners by using a number of questioning techniques that depend on the situation as described in the table on the next page.

Question Type	Description
Overhead question	The Trainer directs the question to the entire class and waits for an answer.
Direct question	The Trainer chooses a specific individual and asks that person the question. **Be careful using this technique!** Let's review a few details. • Adults don't like being "put on the spot." If you do decide to use this technique, warn the class that they will be called on randomly throughout the course to answer questions. • Some Trainers only use this technique if they are sure the person knows the answer. Keep in mind that learners may be afraid to respond in front of their peers and/or supervisors. • Bear in mind that adults have complex lives filled with life issues, challenges, and demands. A learner could be preoccupied with personal thoughts and simply not in the mood to participate. • This technique is a good way to control a talkative or inattentive participant. However, use it sparingly. • When self-study or reading materials are assigned, it is difficult to know for sure if the assignments were completed or understood. Trainers often use direct questions to make sure.

Question Type	Description
Combination direct and overhead question	The Trainer directs the question to the entire class and waits for an answer and calls on a specific individual to answer.
Relay question	The learner asks the question, but the Trainer redirects it to the group. This technique encourages interaction, sharing, and collaboration among participants.
Rhetorical question	The Trainer asks a question that has an obvious answer. For example, you might ask, *"Who wants to break early for lunch?"*

Questioning types

Asking the Question

A Trainer can use a number of questioning techniques (overhead, direct, relay, etc.). However, avoid asking ambiguous or trick questions as well as long or involved questions. As a Trainer, it is important how you ask the question as well as how you respond to your learners.

When asking a question, ask the entire class so no one is singled out. Be careful calling on people unless you are sure they know the answer. This added pressure is not always welcomed. Remember, being placed in a learning mode can be stressful.

Once you ask the question, give learners time to process the information. In Chapter 2, we discussed the variety of ways learners process and recall the material. They might need time to hear the question and formulate their thoughts about what was learned. Ten seconds is typically the accepted time to wait before giving the answer. Those who wish to answer will respond. After a few more seconds, if no response is given, use another questioning technique, or simply answer the question.

Factors to Consider When Asking Questions

In most instances, it's not a good idea to ask questions about content that you have *not* taught. Use your learning objectives to formulate your questions and assessments. They are a good way to measure whether or not the content is understood.

Each classroom will have its own unique personality—sometimes you may encounter a spirited, interactive group that is excited about the subject matter and other times the participants are quiet and unassuming. As you become a skilled Trainer, you will develop the ability to "feel" the personality of the learners in the classroom. If you have an energetic group and you want to provoke and stimulate thinking, you can ask questions about content that has not been taught. Be sure to use this technique only when you believe the audience can handle it and you are positive a knowledge transfer has already occurred.

Responding to Questions

A good Trainer encourages participants to ask questions. Questions actively involve the participants in training, and provide an additional opportunity for everyone to learn by sharing and collaboration. Your response may affect how members of the audience participate throughout the day.

Often, someone will ask a question that is confusing and unclear because of a lack of understanding. When this happens, restate the question as you understand it to verify that your interpretation is correct. Then repeat the question to the entire class.

Using Tips and Techniques to Respond to Learner Questions

Let's review a few tips on effective ways to respond to questions.

- If a learner asks a simple question and then another person asks a stimulating and thought-provoking one, be sure to reward each person. You can say something like, *"Great question, Beverly."* All questions should be treated equally, whether simple or complex.

- If a question is relevant to the topic, answer it immediately. If the question relates to a topic that you will cover later in the session, wait and answer it then. Your goal is to stay on target and avoid straying from your agenda.

- If the question is unrelated to what you are teaching and moves the discussion to a topic that is not part of the agenda, advise the learner that it can be discussed at break or after the session.

- If a student answers a question incorrectly always provide a positive response such as, *"Mya, that was a great effort."* You must always praise and reward the learner for trying. Praise creates a nurturing environment and the student may offer to respond again, even if the answer is incorrect.

- Avoid eye contact if one student is dominating the floor. This non-verbal technique may appear rude, but your goal is to allow everyone to participate. Then, restate the question to allow other students to respond.

- If you use the direct questioning technique to select a specific person to answer a question, use an open hand with the palm facing up. This hand gesture demonstrates that you are open and willing to listen to the answer.

- If you ask a difficult question and there is no immediate response from the class, remain silent and allow the class members time to formulate their thoughts. The rule is to wait ten seconds. If there is no response, use another questioning technique such as next step, leading, or reflexive to provide hints.

- If you find that a learner does not want to participate, appears shy, or is too quiet, pose a question to the entire class, and ask the quiet learner for his or her opinion. For example, you might ask: *"Does that sound right to you, Lorraine?"* Use this technique sparingly; the goal is to promote learning, not to embarrass anyone.

How Responding to Questions Affects Your Credibility?

As a Trainer, your credibility may often be challenged. Your goal is to demonstrate expertise in the course content as best you can. However, Trainers don't know all the answers. Often, we can teach the technology, but because we are not fellow employees, we may not know how to perform the day-to-day functions of a person's job. This disconnection often leaves Trainers at a disadvantage. Keep in mind that we are human beings, too! If you do not know the answer to a question, be honest about it. Do not deliberately mislead the audience by inventing an answer. If you don't know the answer you can redirect the question to the group. You can respond by saying, *"I'm not familiar with this concept. Does anyone know the answer?"* Be honest at all times!

In addition, you can create a "parking lot." At the beginning of class, create a flip chart and write down all questions that need assistance. You can respond with, *"I don't know the answer to your question, so let's place it on the parking lot."* You might be able to research an answer after the session or during a break. Be sure to let the learner know when you will have an answer. For example, *"I will research it and have an answer for you within two days."*

The Parking Lot

If you are unable to answer a question, put it in the parking lot. Be sure to tell your learners when you will have an answer.

Now you know how to encourage interaction and participation using effective questioning techniques. Part of interacting with learners is cultivating active listening skills and not misinterpreting their needs or wants. In addition, while conducting classroom activities, you will need to observe their body language and non-verbal signals to identify their reactions (discussed in Chapter 5).

Developing Active Listening Skills

To respond to learner questions effectively, you need to develop good listening skills. Let's look at ways to develop good listening skills.

Attending skills involve giving your physical attention to your learner. Non-verbal behavior indicates you are paying attention to the speaker. Attending skills include:

- Being interested and showing attention

- Establishing a posture of involvement (don't appear disinterested, distracted, or a lazy listener)

- Using good eye contact

- Developing active body language and mirroring the learner. For example, if a customer (learner) is frustrated about performing a task, it is not appropriate for you to smile. In this example, you should mirror the behavior using a facial expression of empathy and understanding.

Following skills allow the learner to know that you are listening. Minimal encouragement indicates active listening. For example, *"I see, right, really, mm-hmm, go on, oh?"* These short phrases do not interrupt the speaker's flow, but at the same time allow someone to know that you are actively listening.

Silence allows the speaker to proceed at his or her own pace. It allows the speaker time to think about what he or she is going to say. Many listeners talk too much. They may talk as much or even more than the speaker.

Questioning skills (discussed earlier) help to clarify the goals of the conversation. Asking infrequent questions (without interrupting the speaker) is a necessary part of the interaction between you and your learner. Ask closed questions for a simple response. For example, *"Stephanie, did you set up a new password?"* Ask open questions to help you better understand the speaker. For example, *"Rodney, can you explain how you created a new password?"* As the listener, don't direct the conversation. Avoid asking too many questions in succession. Be certain to ask only one question at a time.

Tips on Listening Well

- Be sure to check for understanding! Paraphrase the message. For example, *"Tyler, did I understand you to mean . . . ?"*

- Put your effort into understanding what the learner is saying. Don't think about yourself and your issues.

- Set aside your own urge to speak and control your need to react.

- Avoid listening up or down. Often a Trainer might inadvertently assign value to people based on age, income, position, and gender. Try to listen with mutual respect. Make an effort to eliminate your assumptions about a particular person or group.

- Note discrepancies—the learner says one thing but his or her body language says something else.

- Listen, don't just hear words.

- Observe body language, non-verbal behavior, tone of voice, and gestures. Look for words that express feelings such as happiness, sadness, and frustration.

Posing the right kinds of questions offers the best way to check for understanding. Asking questions is the most effective way to interact with your students. By asking meaningful questions, you can motivate, engage, and interact with your learners. In the next chapter, you will learn how to manage challenging situations and difficult learners.

Managing Challenging Situations and Difficult Learners

Before you can make your training events memorable for your learners, you must learn how to deal with a variety of classroom situations that *might* occur given learners with varying skill levels, personalities, and attitudes. An assortment of individuals will come into your classroom that may disrupt or interfere with the flow of your agenda. There are learners who are: angry, resistant to change, can't keep up because they did not complete the correct prerequisites, talkative, aggressive, preoccupied with using technology (texting, surfing the web, or emailing), not paying attention, typing while you are talking, having sidebar conversations with neighbors, compulsive answerers who prevent others from participating, lost and confused, refuse to participate, demonstrate aggressive body language, ask excessive questions, silent or aloof, know-it-alls, poor communicators, argumentative, closed-minded, or challengers who attempt to upstage you and discredit your knowledge in the class. Challenging situations can disrupt your class, but there are techniques that you can use to dissuade pesky, difficult learners.

Let's take a look at what happened when I faced a group of hostile learners for a new system deployment.

Students walked into my classroom hostile, resistant, and scared. They were ANGRY! The class was designed for a group of account managers to learn a new budgeting software system. I was prepared and ready to do my job. What I didn't know was that the go-live date was in three days and yesterday the employees were told to report to my training class at 9:00 a.m. For the employees, this meant three days to learn a new system, inadequate time to practice, and no real understanding about the transition that was about to occur. Many were panicked and made statements such as *"Will I still have a job?* or *Why didn't they tell us?"* No change management procedures were developed to prepare them for the new deployment.

The angry, uninformed employees took out their frustrations and disappointments on me. During my system overview segment, two participants got into an argument about the proposed new functions and provided their opinions about what the system design "should" do. I logged onto the new system and when the first screen was displayed the learners started complaining about how complicated it looked. Before I could describe the benefits and values, I was constantly interrupted and complaints were openly discussed about the organization and its management style. By 10:00 a.m. it was time for the class to work with the system. Several people refused to participate while the others continued with complaints. One guy became so angry he threw a chair across the floor. I immediately stopped talking and became still. Then the room was suddenly filled with everyone talking and complaining.

As a Trainer, you must always be ready for the unexpected and adapt and react quickly. It's important to manage your classroom and this situation was beyond anything I had ever experienced or witnessed. Therefore, I made the best decision about how to manage this group. At 10:15, I opened the door and dismissed the class.

I contacted the project manager who hired me. When he arrived, along with a few employees, he apologized for the behaviors of their co-workers. I informed him that I was an excellent Trainer but the only way I could work with this bad tempered, hostile, angry group was if security was present. I also stated that this group required intervention that I could not provide, especially as an outside consultant. I did tell him about the complaints and suggestions expressed by his employees and how they felt they were not listened to even though they were performing the functions of the job. After 25 years in the industry, it was the first time I encountered this type of hostility.

The go-live date was rescheduled and I returned to conduct training three weeks later. At this time, I was able to conduct a successful class with cooperative participants.

Working with Challenging Participants

As a Trainer, it's a wonderful day when your class is following the agenda and the classroom is filled with interaction, sharing, collaboration, positive responses, knowledge transference, as well as fun and excitement. However, a few difficult learners can change the mood of your classroom. This can be the toughest part of a Trainer's job and the most time consuming and draining. Remember, you have to be in control—it's your classroom! Difficult learners are simply different, and you must quickly adapt and regroup.

Adult learners bring a wide variety of challenges into the classroom. Each Trainer brings his or her unique personality into the classroom as well. What might be challenging for one person could be an opportunity for another. For example, slow learners offer an opportunity to use my skills as a Trainer and then measure the effectiveness. Each class that you teach will have a unique personality, too. Sometimes you will get a group of vibrant and eager learners. At other times it might seem like you have to beg participants to talk with you.

Let's look at the two groups.

1. Learners' **skill levels** can have a major impact on whether you stay on track with your agenda. If there are slow learners who have not mastered the necessary prerequisites, they slow down the class. Fast learners can become bored. If there are learners in the class with mixed skill levels (novices, beginners, intermediate, or advanced) you have to figure out how to accommodate all of them and make adjustments to the content you are teaching.

2. Learners' **personality styles and attitudes** can be your biggest barrier. In Chapter 2, we discussed how adult learners come into the classroom with the day-to-day issues in their lives. Remember, learner behavior such as resistance, anger, or disruptive behaviors could be symptomatic of other issues present in their lives. Often, they may misplace their feelings (consciously or unconsciously) onto you. For example:

I was teaching a software class to a group of assembly line workers for second shift. The company was moving from a manual process to an automated one. A customized software program was designed for the new initiative. I had repeatedly stressed that students should arrive on time on Thursdays because I would conduct a comprehensive review session, and it was necessary for everyone to be in attendance.

On this particular Thursday, Diane, one of my best students, arrived 35 minutes late. I walked over to her and quietly asked if she had read my email about the rules regarding tardiness. Her body language implied that she was in a very bad mood. She looked at me and said, *"My car broke down, I had to drop the kids off, my youngest is sick with a high fever, I had an argument with my husband, I didn't take a bath this morning, I'm hungry, I have no money, and I'm tired, so please leave me alone."* So, I simply followed her request and walked away. I went to the cafeteria and purchased a coffee and donut. When I returned, she was lying head down on the desk, and she was not participating in the practice exercise. I placed the coffee and donut on her desk and when she looked up at me I placed my hand on her shoulder and told her how proud I was that she came today.

When we dismissed for lunch, I asked if she would join me. We didn't discuss her personal issues. Instead, I talked about a few TV shows I had seen the previous night, and then we went over the topics that she missed earlier in the day. I told her she was one of my best students, I was proud of her progress in my class, and I was glad to have her as a student. At the end of class, I gave her a few dollars to get through the week and told her to continue achieving her goals and being a great mother.

As a Trainer, it is impossible to know what is going on in the personal lives of your students. However, always try to create a nurturing environment as best as you can. Five years later, Diane, my former student, is a systems analyst. She continues to keep in touch with me with a Christmas card each year. She always reminds me of what I did and said to her on that Thursday morning five years ago.

Working with Challenging Skill Levels

Lorraine, the Fast Learner

When appropriate, you can use a fast learner to pair up with another participant. Never let the fast learner dictate the flow of your class. Do the following:

- Suggest that the fast learner, if he or she is willing, pair up with someone who needs help.

- Give additional exercises and set new goals to reduce boredom.

- Involve the fast learner to include him or her in the learning process. When questions are asked, direct a few of them to the fast learner so he or she can participate and interact with the group.

- Make sure the fast learner is functional and doing something; don't permit time-wasting.

Stephanie, the Slow Learner

Use the slow learner as an opportunity to show your teaching skills and to see if you can help him or her make the knowledge transfer. As discussed earlier, adult learners have a decrease in short term memory and processing new information. Do the following:

- Identify if the problem is learning to master the new information. In this case, offer more practice exercises and be sure to reward the improvement immediately.

- Determine whether the problem surfaced because prerequisites haven't been met. In this case, provide individualized instruction. If you determine the slow learner cannot learn the material without taking the prerequisites first, simply ask the person to complete the necessary entry-level requirements, and then reschedule the class.

- If the learner is a true novice, provide lots of practice exercises to help him or her master the new information. Practice is essential to recall the material.

- It's tempting to devote more time to the slow learner, but you must stick to your agenda. Slow or fast learners cannot dictate how you manage your class.

- If someone needs help, ask if he or she would like to be paired with another learner for assistance. Be careful with this! Ask permission first because a learner may not want a co-worker to know that he or she is having difficulties with the topic.

- Offer homework or other resources.

- Offer to stay late or provide your phone number for a later discussion.

- Shorten your breaks and offer to be available.

- Be patient and understanding, even if you have to respond to the same question multiple times.

- Avoid any non-verbal behavior or body language that may indicate your frustration. Your job is to teach! Be patient with your slow learners and remember, you are there to teach.

- Use Step 4 of the training model—"Use the 3R's —Review, Reinforce, and Reward."

Thomas, Nicole, Mya, and Tyler All Arrive With Mixed Skill Levels

When you have learners with a mixture of skills levels (novice, beginner, intermediate, or advanced) watch out—this can be a tough one. You will be able to identify this quickly as you walk around the room to monitor your learners' progress. You can't let the slow or fast learner dictate the flow of your class. You must teach according to the timings on your agenda. If there is a marked disparity of skill levels among learners, you may have to segregate the class (I don't mean physically move people around) and provide slow learners with one set of goals while giving fast learners a different set to keep them motivated. When you have students with mixed skill levels, be sure to keep the pace at the *average* skill level. As a Trainer, you must always be ready for the unexpected and adapt and react quickly.

Working with Challenging Personalities, Behaviors, and Attitudes

When I am in front of a group I am anxious to share all of my knowledge with my learners. However, nothing annoys me more than participants who are talking while I'm instructing, typing while I'm talking, or those who challenge my expertise beyond clarification and curiosity questions. There are numerous other distractions that can impede the flow of your class, but these are my pet peeves. You may often encounter resisters or challengers in your classroom, which can be emotionally draining for the Trainer and the learners who want to learn.

Let's look at a few ways to remedy these challenging situations.

Pearline, the Know-it-all Expert Learner

Eventually, all Trainers will encounter learners who will challenge your knowledge of the subject or attempt to dominate the room by letting everyone know how smart they are. Do the following:

- Instruction is based on collaboration and sharing, so if the person appears to be an expert, listen skillfully and use it as an opportunity to learn something new. You can respond with, *"Pearline, this is great information, and it appears that you know about this topic."* Let the learner share with the group without disrupting the flow and timings on your agenda.

- Don't allow know-it-all learners to take over. During a break you could ask them to monitor their behavior and not to interrupt your instruction. Be sure to use a neutral tone and non-threatening body language.

- Consider asking the know-it-alls to assist other learners because they clearly understand the functionality of the system. Keep in mind that some individuals need and want to be heard so you may have to patronize them to minimize the interruptions. Involve them so they are part of the process.

- Most importantly, if you don't know something and the expert does, don't feel intimated; simply explain to the group that you are unfamiliar with the topic. If necessary, place the question or concern on the "parking lot" and get an answer later.

Clarence, the Talkative or Whispering Learner

A training class can be difficult to manage when the participants all know each other. If one person starts a conversation, it is possible for others to join in and disrupt the class. Training classes are often more productive when people don't know each other and they are held at an outside location. The disruption of talking means that the Trainer will have to take time to repeat the lesson; it can anger other participants who want to learn, and cause you to lose your credibility if you don't mange the talkative learners. Do the following:

- When someone is talking, stand right next to them and continue talking and teaching. It is likely that this invasion of a learner's space will be unwelcome, and he or she will get the hint.

- During your instruction if you hear someone talking, quickly identify the person and say, *"Thomas, is there something you want to share with the group?'*

- If several people are talking, simply stop talking. Silence will force everyone to take heed and then say, *"Matthew, do you need help with something?"*

- Ask a direct question of the offender such as, *"Nicole, do you have a question?"*

- If the learner is particularly obnoxious, ask the class, *"How can we respond to Reggie?"*

- If the learner is aloof or not listening, you could ask, *"Does that make sense to you Mya?"*

- Locate the offender and say, *"Clarence, I don't want you to miss anything so please listen so I don't have to repeat it."*

- During our discussion on questioning, I recommended to avoid putting anyone on the spot. This also applies to the talkative learner. Although it may be tempting to single out someone who is talking during your class, avoid doing this. If the talking is excessive and out of control, slip the person a sticky-note with the words "please be quiet." If this still doesn't work, at break time speak to the person and if necessary ask him or her to leave the class. Remember, it's your classroom and you must maintain control at all times.

- Don't insult your learners by using annoying whistles or noise makers. I attended a class where the Trainer would blow a whistle when talking got out of control. I thought, *"Are we children?"* Learn to control the group and manage the talkative participants using subtle techniques.

Rodney, the Resistant Learner Upset about Change

Most human beings are comfortable with things that are familiar and habitual. Change is welcomed by learners who view it as an opportunity to enhance job skills, but it often causes anxiety for others. New system functionality can cause employees to feel threatened about their job security, performance, and future. Keep in mind that change can cause great apprehension in adult learners.

If the organization has implemented a change management strategy for the new training intervention such as company newsletter announcements, kick-off meetings (with balloons, refreshments, cheerleaders, executive sponsors who discuss the benefits), lunch and learn sessions, system demonstrations, or posters displayed around the office, your job will be easier. However, if there is no change management in place, you may be the person with whom learners take out their frustrations. If this is the case, allow a few minutes to let them blow off steam. Simply listen to their stories and complaints without judgment and maintain your agenda. Do the following:

- During a new system implementation, you may hear the following comments: *"Why do we have to change?" "I don't like this?" "This makes no sense."* You can describe the new features and functions of the system and emphasize the benefits and values of learning the new information.

- Present new information by connecting it with something the learner probably already knows. Comparison learning relates the new topic to a known topic. For example:

  ✓ In the old system, you entered a purchase order record as . . .
  ✓ *In the new system, you will enter a purchase order record as . . .*

  ✓ In the old system, there were 10 steps to enter a purchase order.
  ✓ *In the new system, there are 5 steps to enter a purchase order.*

- Ask the class what they liked about their old software and what they don't like about the new system. Although there is not much you can do, the goal is to allow them to discuss their frustrations. Perhaps they will discover the benefits on their own during your presentation.

- To reduce the anxiety and pressure about learning the new system, explain that there is always a learning curve and it will take some time to master the new material.

Carl, the Angry Learner

Remember, adult learners arrive with a variety of issues (such as work, home, health, or finances) and may displace their anger onto you. In instances where the behavior is aggressive and unmanageable, it may be necessary to contact a manager to request that a learner leave your class. Remember, you must always be in control and keep in mind that an angry learner can poison your class. Do the following:

- Avoid arguing—As tempting as it might be, don't get into a discussion or argument with an angry learner. Don't mirror their behavior. Your goal is to create a safe and non-threatening learning environment even when you have angry participants.

- Listen carefully—Whatever angry learners have to say needs attention; after all, they may have some good ideas. In some instances, people who work with a system on a day-to-day basis know a lot about the functions. They can offer useful suggestions that no one has considered. You may be able to make these suggestions to a manger.

- Don't take it personally—Positive responses such as *"I understand your frustration"* or *"I think I see your point"* are helpful. Use body language and facial expressions that demonstrate your concern for their issues.

Elizabeth, the Shy and Quiet Learner

Keep in mind that learners might not be in the mood to participate, depending on issues in their lives. Some people are simply quiet and rarely ask questions or participate in group discussions. Do the following:

- Avoid confusing a quiet learner with his or her ability to learn the content. During Step 3 of the training model, you will get the opportunity to measure performance during practice exercises.

- Use a more aggressive approach and ask a direct question: *"Elizabeth, how do you delete a purchase order?"* Be careful not to put anyone on the spot and use this method in good taste.

- Try using games or team tasks to get them involved.

- See the learner during break to determine if something is wrong, especially if the behavior is extreme.

- Ask easy questions, then more complex ones that require group discussion and interaction.

- Pose a direct question targeting specific learners. For example, you could ask, *"What are the six steps to create a purchase order?"* Then you could ask, *"Mya, what's step 1; Tyler, what's step 2; Elizabeth, what's step 3; Damon, what's step 4; Lurie, what's step 5; Nikki, what's the final step?"* This technique allows you to include a group of learners, including the quiet learner. Start by asking a few closed questions and then build up to open questions that require more discussion and critical thinking skills.

Sue, the Chronic Latecomer

Always start your class on time! Don't penalize the participants who arrived on time eager and ready to learn. Be sure to thank participants who are on time. As stated previously, you must maintain your agenda and class timings. Interruptions will completely disrupt the flow of your class.

It's always difficult for a Trainer to begin class only to stop and repeat what was just said for latecomers. Therefore, establishing ground rules from the beginning is essential. Post the start and end times on the white board for breaks and lunch. If there is a chronic abuser, speak to him or her privately. Don't interrupt your class for latecomers; instead, accommodate them as best you can by working with them individually, or assign catch-up to a neighbor. In addition, don't make assumptions about a latecomer. It is quite possible that there were transportation, day care, or traffic issues. Or, perhaps the student feels marginalized for some reason and is acting out.

Jackie, the Preoccupied 'Busy Bee' Learner

Our technology-driven society forces all of us to multi-task in our work and personal lives. Personal digital assistants (PDAs), smart phones, and other personal communication devices allow us to email, twitter, and text in order to access the latest information on demand. These multi-tasking activities can be disruptive to your class for several reasons. Even though these devices can be operated quietly, learners are distracted by them and you might have to repeat the information. In addition, the busy bee might try to catch up by disturbing the flow of his or her neighbor's progress by asking questions. Do the following:

- Reinforce the timings on the agenda. Announce that breaks and lunch are the times when devices should be used.

- Identify those learners using the Internet or other devices during your instruction. Use the u-shaped room layout (discussed in Chapter 8) to zoom and scan the room.

- Set and maintain the ground rules consistently during the class.

John, the Dominating Compulsive Answerer

A compulsive answerer is a learner who monopolizes a classroom discussion and might prevent the involvement of other participants. Do the following:

- Respond to the compulsive answerer with *"Great answer."* Then, redirect the question to the group and ask another learner, *"Jeff what do you think?"*

- Acknowledge the compulsive answerer by nodding, but make eye contact with the group by scanning the room to locate someone else who appears to know the answer.

- When the compulsive answerer speaks or answers yet another question, ask another participant, *"So Stephanie, do you agree with John's answer?"* Then, ask another learner, *"Lorraine, do you agree with John and Stephanie?"* After a few more responses confirm the answer.

Shirley, the Confused Learner

You must set the stage for the confused learner. These learners respond to you if you make them feel welcomed, comfortable, and nurtured. During the review of my training class evaluations, a student wrote *"Thanks for not making me feel stupid."* Learners need to feel that it is O.K. to ask you a question. No one wants to feel left out, judged, or embarrassed about seeking help. Even if you have to respond to a question multiple times, be sure that your body language is inviting and encouraging. Keep in mind that at some point, we are all novices at something.

Does confused, lost, or slow mean the same thing? Absolutely not! A learner could be *confused* because the instruction is unclear and clarification is needed. A learner could be *slow,* because he or she is a novice and more practice is needed to master the material. A learner could be *lost* because he or she is preoccupied with personal issues, not paying attention, or multi-tasking. A learner could be *confused, lost,* and *slow* if he or she lacks the necessary prerequisites. Lacking the correct entry level skills will most likely result in a failed training intervention. Poor results can lead to self-esteem issues with the learner and increased feelings of fear and anxiety. Nevertheless, whatever the problem—whether a learner is confused, lost, or slow, you will need to identify the problem and make adjustments quickly.

Listed below are a few statements I have heard during my training career.

"I'm so slow on computers."
"I'm afraid of this new software."
"My supervisor is here, so please don't call on me."
"Don't embarrass me if I mess up."
"I'm so stressed about this class."
"Will we be tested?"
"Please don't call on me."

When I encounter confused learners, I automatically know that practice, reinforcement, praise, and reward are necessary. I also know that individual or customized instruction might be necessary. As an incentive for correct (or any) responses, I play Q & A games and offer prizes. You can minimize anxiety and fear by making learning fun and interactive. Do the following:

• Discuss the value and benefit of the new learning.

• Convince them that your job is to help.

• Offer one-on-one instruction.

• Praise and reward consistently for any accomplishment.

• Provide positive feedback and let them express their frustration.

• Monitor progress during Step 3 of the training model (while learners practice). It's a good idea to check on them every few minutes. You can stand over their shoulders and watch them without interfering. If they are doing a good job, tell them. Reward learners as often as you can. If you notice something is incorrect or their body language demonstrates confusion, intercept immediately.

• Offer to help during breaks and provide homework.

Handling Environmental Situations

Consider the needs of busy adult learners and be sure to create a comfortable environment. This includes:

• Create a comfortable workspace. Make sure the desks are not crammed together. Check for exposed cords, network cables, or electrical wires located where people can trip over them.

• Ensure that the room is well lit and appropriate for visual aids.

• Ventilate the room adequately.

• Adjust the thermostat if necessary. If the room is too hot or too cold, learners will instantly complain. The room temperature can also depend on your audience—some people prefer the room to be colder than normal.

- Avoid having an overcrowded training room. The number of participants enrolled is typically up to the client, but request 12-to-15 students maximum per class.

- Make sure each person has a working computer.

- Minimize noise distractions. Try to use a training room that is far away from the lunch room, coffee or water machine, or a hallway with high traffic.

If the client permits it, use the u-shaped format to lay out your training room (discussed in Chapter 8) to make it appealing and functional for you. This means you might need to rearrange the furniture if it is feasible.

Launching the Training Day

Before you can make your training events memorable for your learners, make sure you are organized, prepared, and ready with the right tools for a successful day.

Let's take a look at a training class that I attended.

I arrived early for a Dreamweaver class. I did a series of readings before the class and even downloaded the software so I could have a head start. I was eager and highly motivated to learn this new technology. In addition, it cost a few hundred dollars to take the course. When I walked into room 702, the PCs were not turned on and there were empty water bottles, and soda cans left on several desks as well as papers. The room was not appealing at all. The class was scheduled to start at 9:00. The Trainer arrived at 8:55. He took 20 minutes to load his machine because he told us that he needed to upgrade a new version of the software. He then determined that the projector needed a light bulb. Once this was fixed, the participants turned on their PCs to learn that four didn't work. He called the technical department but was told no one could arrive until 11:30, which meant several people had to double-up on computers.

Therefore, paying customers would not receive hands-on experience. The class actually started at 10:47. No welcome, introduction, or learning objectives were offered. To make matters worse, no learning materials or reference documents were provided. Several people left at break and reported the Trainer to the school. I was angry and disappointed because I wasted an entire day.

What's My Style?

I like to make my training room comfortable and conducive to learning. I always, always, always, arrive early to handle any unexpected situations. Remember, adults may come into your classroom with the day-to-day stresses in their lives and learning something new may be difficult to focus on. Dealing with unforeseen obstacles like network problems, poor lighting, a dirty room, or passwords that don't work will probably not be welcomed.

About My Room Preferences

I never let learners walk into an empty room. I make sure there is a PowerPoint presentation on the overhead projector when they walk in. The title slide will display something like, *"Welcome to the Purchase Order Entry Course"* along with my name as the Trainer. I like to make the room warm and comfortable. I will also have soft music playing and spray an aromatherapy fragrance.

In addition to the PowerPoint presentation displayed on the LCD projector, my room will be decorated with visual aids. I like to have blown up posters of the content I plan to cover. For example, if I'm teaching about purchase orders, I might have a completed purchase order with callouts to provide a visual of the topic. This allows me to refer back to the poster periodically during training. Visual aids are a perfect way to capture and maintain attention.

I also hang two posters. One is called *"Entering the Learning Zone"* and the second is an illustration of my *"four-step training model"* (discussed in Chapter 4). As part of my introduction, I like to explain the model and process that I will use throughout the day. If the client has budgeted for job aids, I like to issue them when the learners walk in, especially if they are glossy and laminated. If there are takeaways such as cups, paper, mouse pads, or pencils, I like to give them out at the beginning of class. Everyone likes "takeaways."

What Happens When Students Arrive?

When the students arrive, I greet them with a warm welcome and tell them I am glad they chose my class. I even assure them it will be fun. I ask them to sign in and offer them refreshments. I like to wear my training button on my blazer that reads **"Students Come First."**

I give them a name tent. I think it is important to know your learners and interact with them using their names. Names are important because I like to play games, what-if scenarios, and team tasks when I conduct my Q&A session. For example, I might say, *"It takes 6 steps to create a purchase order. Lurie, explain step 1; Matthew, explain step 2; Michael. explain step 3."* If I don't know their names, I can't interact with them.

Once the sign-in sheet and name tent are completed, I ask them to use the sticky-note that is stuck to their chairs to write down their learning goals for the day. Then, I collect them and place them on a flip chart in the front of the room. During the course introduction, I will read each sticky-note so all expectations are clearly identified.

At lunch time, I like to give out a sticky-note again. This time, learners can secretly tell me how they feel training and the pace is going. Depending on the response, I may need to make a few adjustments.

What about the Room Layout?

Some organizations still set up their classrooms using a traditional elementary school floor plan with desks in rows while the Trainer stands in front. This dictates the traditional leader-follower style. Instead, if the client will permit it, I like to set up the training room in a non-traditional method using a u-shaped format.

I like this style because I am in the center of everything. I can *zoom, scan,* and *find* everyone in the room within a few seconds. It also allows me quickly to see the learners' body language and non-verbal reactions during practice exercises. Most importantly, I can quickly identify if someone is not keeping up with the pace of the class. For example, if I'm walking through a process and I'm on step 7 and a student is on step 4, I will know immediately.

The u-shaped format also allows me to stand in the middle or back of the room with the ability to view the students' progress and provide individualized attention. I am also able to move freely from person-to-person.

Prior to the Training Day

Prep, prep, prep, and more prep. Never stand in front of an audience if you have not "prepped" (prepared) first. Preparation is the key! Be sure to rehearse your presentation prior to delivery. Your credibility is on the line and your learners are looking at you as the SME. You are the role model and the person responsible for the group, so be sure you are prepared. Make sure you have an agenda with the scheduled timings for each topic. Be sure to allocate time for questions, system issues, or unexpected events involving lighting or climate.

Managing the Training Day

As a Trainer, there is nothing more frustrating than a class that does not go smoothly. This can range from learners who come in late and interrupt the flow of your class to technical difficulties. Your job is to stay on schedule, keep things moving smoothly, and create a positive experience for your learners. The process for the day includes the following:

1.   Prepare the classroom.

2. Greet the participants.

3. Introduce the course.

4. State overall course learning objectives.

5. Discuss the agenda and schedule.

6. Deliver training using the four-step training model for each unit of instruction.

7. End the course.

8. Complete the course evaluation.

Preparing the Classroom

Arrive early enough at the training site to prepare your classroom. Be sure to arrive before the participants so you can begin on time. Make sure the PCs, projector, or any media are working properly. The training materials, handouts, or job aids should be printed and ready for distribution. Make sure you have a contact person if technical difficulties occur. Most importantly, if you are unfamiliar with the room or facility, it's a good idea to visit it ahead of time. Before you begin, complete this checklist:

• Training script, presentation, notes, etc.

• Class roster

• Sign-in sheet

• Training materials, job aids, handouts, etc.

• Name tents

• Check PCs

• Check projector and always have extra light bulbs

• Check any other technical equipment such as printers, scanners, network cables, etc.

• Confirm room assignment

- End-of-day evaluations

- Assessments, if applicable

- Sticky-notes, paper, pens, and markers

- Visual aids if applicable

- Contact person for technical issues

- Training coordinator's name and phone number

- Check the room temperature

- Parking lot to capture questions

Greeting the Participants

As your learners walk into the room, begin developing a rapport with a warm smile and greeting. Depending on your company, some Trainers play music and have food available to create an inviting environment. When you begin the class, greet the participants enthusiastically. First impressions are so important. You can establish your presence and take control of the course with a strong first impression.

Introducing the Course

Introduce the course with a friendly and inviting facial expression. There are a few things that you can do for the introduction:

- Learner introduction—Have each learner provide an introduction. For example, name, department, job responsibilities, and goals for taking the course. Other introductory exercises include:

  o Sharing an embarrassing moment

  o Describing a favorite activity

  o Getting to know you activities (true and false statements, questioning games, etc.)

  o Conducting a brainstorming session – Why are you here?

- Trainer introduction—Introduce yourself and provide a brief overview of your training experience, credentials (if necessary), and background. This helps to establish credibility.

- Icebreaker—Complete an icebreaker or warm-up exercise to reduce tension, get people involved, open up communication, and have some fun.

- Learning goals—Pass out a sticky-note and ask each person to write down their learning goals. Be sure to collect this immediately. Place them on the flip chart to see if there are similarities and to identify expectations.

Stating the Overall Course Learning Objectives

Earlier in our discussion, we learned about using the four-step training model and the importance of stating the learning objectives. Be sure to discuss:

- Course objectives: describe the overall objectives for the entire course.

- Chapter and topic objectives: describe what you will teach for each chapter and topic. The most important time to motivate the class is at the beginning of a new chapter and topic. This is the best opportunity for the Trainer to capture the learner's interest in the concept about to be presented.

When you state the overall course learning objectives, you set the expectations for the day. You can help participants identify common ground between their personal objectives and the topics that will be covered during the day.

Describing the Agenda, Timings, and Schedule

The agenda should be prepared with specific time allocated for each event. Class timings provide structure to the day by breaking the class into manageable components. If participants know the schedule, they can plan for breaks, lunch, and the end of the day. Estimate the time needed to teach each chapter. For example:

Topic	Time Allocated
Welcome	8:00 – 8:30
Chapter 1	9:00 – 10:15
Break	10:16 – 10:30
Chapter 2	10:31 – 11:30
Lunch	11:30 – 12:30
Chapter 3	12:35 – 1:15
Chapter 4	1:16 – 2:15
Break	2:15 – 2:30
Chapter 5	2:31 – 3:30
Recap, Review, Final Q & A	3:31 – 4:00
Assessment	4:00 – 4:30
Complete Evaluation	4:30 – 4:45
Wrap-up and Dismiss Class	4:46 - 5:00

Agenda topics

You will need to monitor your progress as you teach to check whether you are behind or ahead of your lesson plan. A Trainer may need to make adjustments. Be sure to inform your learners of the time scheduled for breaks, lunch, and the end-of-day.

What about Latecomers?

Always start your class on time! Stay true to your agenda. Minimize the interruptions caused by latecomers who disrupt the flow of your class. Post the start and end times on the white board for breaks and lunch. If there is a chronic abuser, speak to him or her privately.

Housekeeping Rules

Be sure to explain the classroom housekeeping rules regarding recycling, the lunch room, and the location of rest rooms and vending machines. You will probably encounter a few multi-taskers who will try to check their email, cell phone, blackberry, or pagers during class. Inform learners to be respectful of their classmates. Ask them to leave quietly if it is absolutely necessary for them to return a phone call, email, or text message. For example, I taught a software class at a hospital for a group of neonatal nurses who were on call and whenever they received a page, they quietly left the room. When the class is over, make sure each person discards or recycles any papers, food, water, or soda cans as needed. Be considerate and avoid leaving the room in disarray so the next Trainer will not have to clean up behind you.

Being Prepared for the Unexpected

As a Trainer, you must always be ready to adjust and adapt quickly to unexpected events as described below.

What happens if the projector bulb burns out in the middle of a presentation? Always have extra projector bulbs on hand and alternate teaching aids available if you can't get the projector working. You may not have the luxury to cancel the class, so always be prepared to use an alternative method. For example, perhaps you could dictate the steps or process without learners actually seeing what happens via the projector.

What happens if some computers don't work? In instances where the computers are not working or there are network problems, sometimes it is tempting to cancel and reschedule the class. As an alternative, if you have one working computer, you can demonstrate the system functions to the class while your learners take notes and watch your presentation. If only a few computers are working, an alternative is to have learners double up or crowd around the working computer. Avoid wasting time trying to troubleshoot a problem, always find a quick solution and adapt. If all the computers suddenly go down, you can use the screen facsimiles provided in the training manual to walk your learners through the process. Of course, this is not the ideal situation, but it is certainly better than canceling the class.

What happens if there is a fire drill or some other issue? When you walk into any organization, always learn the exit routes and emergency procedures. As the Trainer, you must always be prepared for the unexpected.

Delivering Training Using the Four-step Training Model for Instruction

As discussed in detail in Chapter 4, use the four-step training model for each unit of instruction you are teaching. For example, perhaps you are teaching a group of human resource administrators how to use a new financial software system consisting of the following five topics:

1. Creating a requisition

2. Entering the purchase order

3. Receiving goods into inventory

4. Generating an invoice

5. Performing a search query

Each of these five topics represents a unit of instruction. Therefore, you would use the four-step training model five times to cover each topic.

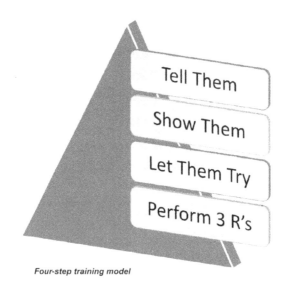

Four-step training model

Step 1—Tell Them (Motivate)—Motivate learners by telling them what they will learn at the beginning of each chapter and topic.

Step 2—Show Them (Instruct)—Instruct learners by defining, explaining, and demonstrating.

Step 3—Let Them Try (Practice)—Practice, practice, practice! Reinforce what you taught them with practice and repetition.

Step 4—Perform the 3 R's—Review, recap, and reward your learners.

Pacing Your Day

Be sure to have your agenda handy so you can keep abreast of how your day is progressing. There may be times when you are right on schedule and the class is going as planned. However, there are varieties of reasons that may take you away from the agenda. This can include accommodating slow learners, answering long and involved questions, holding discussions that strayed to a different topic unrelated to what you are teaching, resolving system issues, or dealing with learners who return from break too late.

Let's look at how to manage these incidents.

When You Are Behind Schedule

If you cannot cover all the items on your agenda, you will need to quickly adjust your schedule and prioritize which topics to exclude. Be sure to include all topics that the learner must know before leaving. You may suggest staying late or meeting learners on another day if the information is critical for their development. If you weren't able to cover certain topics, provide the necessary reading material or resources and suggest that learners cover these topics independently.

When You Are Ahead of Schedule

This is always a Trainer's ideal situation. When you are ahead of schedule, allow for more practice, introduce new topics, or provide more Q & A. You should always have additional topics in your lesson plans.

What about Breaks?

Breaks give the Trainer and learner an opportunity to relax. In instances where there is so much information (data overload), breaks allow learners the opportunity to absorb the details. Typically, a 15-minute break is given after 2.5 hours of instruction. Watch learners' body language and when necessary, allow them to get up and take a quick stretch to relax without leaving the room.

It can be difficult for some learners to return on time from break, so be sure to announce the exact time they should return. Trainers must always resume the class on time, even if the learners return late. It is your responsibility to follow the agenda and learners must be accountable for their actions. You must set the expectation and reinforce it for each break that you offer. In addition, never announce a break and then ask a question. When learners hear the word "break," their thinking automatically shuts down.

What about Energy Levels?

Most adults experience their highest energy levels in the morning. This is the best time to learn and apply new information. Typically, after lunch (or after eating a heavy meal) a slowdown occurs. I prefer morning classes to afternoon classes. If necessary, offer frequent breaks and lead a few stretching exercises to break up the monotony of the training day. You can also use instructional methods such as team tasks, games, humor, and scenarios that provide movement and interaction.

Completing the Training Day

Conclude your course by summarizing the main points and providing a comprehensive review of each chapter. I like to summarize the course by conducting a Q & A session. Then, review the learning objectives as proof of teaching the content.

Evaluation of Learner Outcomes

Evaluation measures the effectiveness of the training program and determines the impact of the instruction. Evaluation tells us whether or not the training instructional goals were met. For example:

"Was there a transfer of learning?"
"Were the learning objectives achieved?
"Can learners perform the tasks learned during training?"
"Did learning barriers exist?"

The Kirkpatrick Model[9] is a four-level evaluation model used to assess the effectiveness of a training program. The four levels of the model essentially measure:

- Level 1 (reaction)—Participants provide feedback about the training they received. Instruments used for level one can include an end-of-the course questionnaire, course evaluation (sometimes called a "smile sheet"), or focus groups.

- Level 2 (learning)—Indicates an increase in knowledge or skill. Instruments used for level two can include pre- and post-skill discoveries, classroom assignments, demonstration of a task (hands-on exercises), question/answer, or problem-solving cases.

- Level 3 (behavior)—Indicates a change in the learner's behavior and capability improvement. Instruments used for level three can include on-the-job observations, self-assessment, or reports from customers, peers, and the participant's manager.

- Level 4 (results)—Indicates the effects on the business or environment resulting from the learner's performance. Instruments used for level four include financial/sales reports, quality inspections, and management reports.

Choosing an evaluation level depends on the organization and its funding sources. However, most firms typically use levels one and two. Even if the client does not recommend doing a level one evaluation using the smile sheet, it is beneficial. You should understand how your training session was perceived and make adjustments if necessary. Let's look at an example of a course evaluation on the next page.

Conducting the Course Evaluation

It is important to evaluate the effectiveness of your training and determine if you exhibit any deficiencies of which you are unaware. The end of the course evaluation (or smile sheet) is most often used. The smile sheet allows

---

9   Kirkpatrick Donald and Kirkpatrick James, *Evaluating Training Programs: The Four Levels Third Edition*, Berrett-Koehler Publishing, 2006

learners to express their feelings about the training and allows *Trainers* to receive feedback about the course content, delivery, and facilities.

Give Us Your Thoughts	Strongly Agree	Agree	Neutral	Disagree	Strongly Disagree
The class met the stated learning objectives.	5	4	3	2	1
I found the information in the class to be useful.	5	4	3	2	1
The hands-on exercises were useful in learning the new system as it to pertain to my job.	5	4	3	2	1
The manual was easy to read and use.	5	4	3	2	1
I can use the course materials in the future as a reference.	5	4	3	2	1
The instructor has a command of the subject matter.	5	4	3	2	1
The instructor's presentation was clear and well organized.	5	4	3	2	1
The instructor allowed opportunities for questions and discussion.	5	4	3	2	1
I would rate the instructor as excellent.	5	4	3	2	1
Adequate time was provided for questions.	5	4	3	2	1
The curriculum content was organized and easy to follow.	5	4	3	2	1
Overall, I was satisfied with the training class (this includes both the documentation and instruction).	5	4	3	2	1

As a Trainer, use this feedback to learn what you did well and if any improvements are needed. Be sure to allow adequate time for the evaluation process.

Final Wrap-up

Always thank your students for coming! Until your learners leave the classroom for good, continue to create an environment that is warm and inviting until the end. While my learners are completing the end-of-course evaluations, I put on soft music and re-open my original PowerPoint presentation. My final slide will say, "Thank You for Coming."

Be sure to encourage learners to use additional resources such as online help, help desk assistance, or web site. If post training is offered (drop-in training room, floor support, WBT or virtual training deliveries, FAQs, or training videos), be sure to provide the details.

Now that you understand the importance of planning your day, preparing for the training delivery, and managing the daily events, you will be able to create a successful educational experience for your learners. In the next chapter, you will hear from professionals in the training industry.

ADVICE FROM TRAINING PROFESSIONALS

I've had the pleasure to work in an industry that I love. I have also had the privilege to work with some outstanding training professionals and colleagues over the past 25 years. These are individuals who I have watched deliver training and were extremely impressive. Each Trainer is passionate about the craft of training and possesses all the characteristics that I believe make excellent Trainers.

What Attributes Make a Good Trainer?

The following list provide a few examples:

- Demonstrates excellent presentation and delivery skills

- Engages, captures, and motivates learners

- Checks for understanding by using a variety of questioning techniques

- Shows care and concern for the learning population

- Accommodates learners by going the extra mile to make sure they understand the content (stay late, offer assistance on their own time--lunch, breaks, or after hours)

- Ensures positive learning outcomes and knowledge transfer

- Encourages participation and promotes independence

- Makes learning fun rather than serious and overbearing. Provides action, interactive learning experiences

- Creates a non-threatening and non-competitive atmosphere

- Respects learners

- States learning objectives and explains why

- Juggles different skill levels

- Provides clear, concise, and accurate instruction

- Manages time

- Listens well

- Offers patience and understanding

- Uses examples and analogies from the learner's world

- Understands adult learning theory and principles

In this chapter, you will read the opinions of eight training professionals. I asked each Trainer to respond to one of the following questions:

1. What attributes make you a good Trainer?

2. How do you connect with your learners?

3. As a Trainer, how do you create a memorable training experience for your learners?

Let's read their thoughts and opinions.

Charlotte Cager
B.S. Education, CTT+ (Certified Technical Trainer Plus)
President, Teachable Moments Consulting, Inc.
Software Training Specialist/Training Manager

As a Trainer, how do you create a memorable training experience for your learners?
I've learned throughout my many years of training it is most vital as an instructor to be patient with learners throughout the learning and training process. As a training professional, I am a disservice to any learner if I'm rushing them through the learning and training process. I must always be aware that every learner learns at a different pace, which is indicative of their learning style. Being patient with a learner sends many positive messages. Being patient with a learner sends a message that: I am concerned with their learning and understanding of the delivered material; I am aware that learning the new material may be challenging to perceive; and, retaining the new material in order to be successful with it can be a major challenge as well. It warms my heart whenever a learner tells me I was patient with them throughout the learning and training process. However, the ultimate reward is the learner can then experience a teachable moment. What a wonderful moment for any Trainer!

How do you connect with your learners?
I use the icebreaker to get to know the participants and have them share information about themselves. During training I use analogies, overhead and direct questioning, and add a dash of humor to draw them into the training. I try to use a conversational style of training in which the learners are fully engaged. I also like incorporating games and break-out sessions. I'm always in tune with my learners by watching for cues to help me determine their learning style so I can train them based upon their style.

What attributes make you a good Trainer?
Passion, patience, honesty, knowledge and experience, respectively.

Adel Etayem
MCP, MOS, CAN
Training and Application Development Manager

As a Trainer, how do you create a memorable training experience for your learners?
A memorable training experience occurs when a Trainer consistently demonstrates a passion, vocation, and aptitude for training. A great Trainer adapts quickly under any circumstances, proactively assessing and addressing the immediate changing needs of the audience. One achieves a memorable training experience by empowering the participants. This means they have learned the material and can effectively train others. The greatest compliment a Trainer can receive is to have the essence of the training duplicated.

I taught a training class to a group of frightened non-technical participants. Their continued employment depended on mastering the material to perform day-to-day operations. The key was to provide a relaxed environment in which to train them and build their confidence. This was done by incorporating basic skills with the course objectives and changing the pace as needed. Humor helped build a relaxed environment. Here was a situation where less was more. I watched the audience for non-verbal queues, such as exhaustion, confusion and most importantly, I listened. At the end, they overcame their fear of technology, embraced it, learned it, and felt empowered to share what they learned with co-workers. Each participant told me that it was a great and memorable learning experience.

Ellen Lehnert
PMP, MCT, MCITP
Consultant/Trainer, Lehnert CS, LLC
Microsoft Project & Microsoft Project Server

As a Trainer, how do you create a memorable training experience for your learners?
Memorable training occurs when the student walks out of class with the
reason they came to learn fulfilled. In other words, each student comes to
a class for various reasons. They are all trying to increase knowledge, but
they all have different goals and learning styles for what they are seeking
to learn. Seeking out what each student wants to learn and delivering it
in a manner so they can absorb the information will create a memorable
experience.

Most of the classes I teach are software and project management related.
At the beginning of each class I ask each student *"If you could walk out of class
with one piece of information, what would that be?"* I write down each student's
response and then make sure I answer that one burning question for
each person. When I present the topic a student requested, I refer to that
student so they know this was their hot button. During a training class I
also try to inject humor, relate complicated concepts to simple examples,
and create memory triggers. Allowing the student to feel comfortable
enough to ask questions when they have them and trying to answer them
fully will also increase the learning experience.

What attributes make you a good Trainer?
* I try to be in tune to the body language and facial expressions of
  the students. Silence could mean that I lost them. For example,
  are they paying attention or doing email?

* I don't allow students to hijack my class and take it in their own
  direction. Instead, their questions are always taken off line. Many
  times I have suggested that someone eat lunch with me so they
  could get their questions answered.

* I try to establish what would make an individual feel that their
  time was well spent and then fulfill that need.

* I am very comfortable with the material that I present. If someone
  asks me something I always have an answer, no matter how

obscure the question. This creates a comfort level when students know that I can handle any question at any level. I also believe that confidence is conveyed to the students as well.

- I know the material well enough so that I don't get rattled when something goes wrong.

How do you connect with your learners?
- I try to use humor to connect with my students. If they say something funny, I make them feel good about what they said. One thing that I often say in my classes when I have a lot of students, is that if they want to make me crazy, they should move around so I wouldn't know anyone's name. One class actually moved the name tags around just to play a joke on me. I congratulated them for their effort.

- In another class I said to someone who asked a good question, "You win a gold star." The student said "Where is it?" I happened to have some in my laptop bag and put one on her book. After that the class was competing for gold stars. Sometimes I purchase odd prizes at the dollar store and give them out for good questions or good answers. Ask a good question and you win Groucho glasses.

- I also let my clients know that I understand their unique situations and could be a source of help to them. I encourage them to feel comfortable about asking questions and let them know when it was a good question or say, *"I'm glad you asked that."*

- Students do not like instructors who read the book to them. I know my material well enough that I can talk it without reading it to the class. As the Trainer, I have more credibility.

Thelma Reed
M. S., Management, Human Resource Development
Instructional Designer, Trainer, Director, Training and Organizational
Development

What attributes make you a good Trainer?
As professional Trainers I think we all recognize the importance of standard attributes such as subject matter expertise, good communication skills, organization skills, patience, and being a good listener. However, there are other attributes that I feel are critical to creating a rewarding learning experience.

- Adaptability/flexibility—A good Trainer is able to adapt to diverse groups with different personalities and learning styles, recognizing that learners have varying degrees of aptitude and understanding.

- Results focused—A good Trainer confirms that the training intervention meets not only the needs of the learner, but also has overall value for the organization. The Trainer must ensure that training goals are linked to business objectives.

- Bridging the gap between old and new methods of learning—A good Trainer will know how and when to blend traditional standards of learning with new ideas, approaches, and strategies.

- Encouraging and supportive attitude—Good Trainers are passionate about what they do and how they do it. They are approachable and enthusiastic, having the ability to motivate and engage the group. They must also be able to manage, persuade, and influence.

- Integrity—Good Trainers do not cross the boundaries of professionalism. They have confidence, use good judgment, and display a positive attitude.

How do you connect with your learners?
The adult learner is dealing with the stressors of work and life issues. They need to feel that the training is relevant and will provide them with immediate solutions to "real" problems. Otherwise, they will easily become bored and disconnect from the training. Therefore, my strategy for connecting with my learners includes:

- Creating a fun, stimulating, thought-provoking learning environment.

- Enabling learning by setting appropriate challenges that are achievable.

- Developing a climate of trust where constructive feedback is acceptable.

- Designing team-centered activities and scenarios that allow each learner to utilize their abilities, skills, and experiences to solve team problems.

As a Trainer, how do you create a memorable training experience for your learners?
Creating memorable training occurs when the overall impact of the training intervention lends itself to solving problems and creating opportunities for growth. For example:

1. The learner demonstrates measurable improvement and high levels of proficiency in both technical and people skills as a result of the training.

2. Effective communications is recognized; there is a positive impact on relationships and cross-pollination of information across departmental lines.

3. There is an enhancement or modification of a process, procedure, or service; quantifiable results are evident.

4. Training objectives and business objectives are linked and in sync.

5. The actual training experience is most memorable when it is engaging and designed utilizing a combination of blended solutions (when possible). The training also should be clear, concise, and relevant, with well-defined objectives. The training is most effective when it meets the "specific and immediate" needs of the adult learner.

Beverly Rico
M.A. Training and Development
Instructional Designer, Software Training Specialist, Adjunct College Instructor

What attributes make you a good Trainer?
The attributes that make me a good Trainer are: I am always willing to change or adapt to the classroom, I always have a plan B, I establish control right away in the classroom, I use a lesson plan for each course I teach, and try to establish an amiable relationship with my class so that they know I am approachable when they are confused or have questions.

How do you connect with your learners?
I like to involve my class when I am training. I let them contribute their experiences with the subject being taught and incorporate their experiences with the learning objectives. I also answer the "What's In It for Me?" (WIIFM) question to offset resistance and gain training acceptance. I also practice a lot before training; the more comfortable I am with the course content, the more professional I will appear. Adult learners like to know they can trust the Trainer's knowledge and they can really learn something from the training.

Mary Stearns Sgarioto
B.A. English, MFA (Master of Fine Arts) English,
M.Div (Master of Divinity)
Trainer, Technical Writer/Editor, Adjunct College Instructor

What attributes make you a good Trainer?
Building trust is crucial to fostering an ideal learning environment. One of the secrets of good teaching is that the best learning occurs when the instructor relates to a student as a human being and draws out the answer the student already knows.

First, showing a student that he or she can overcome challenges—because the instructor and others have—and that there are easy ways to learn and remember material encourages trust and makes a student more open to learning. While instructors know they are experts—and they must have confidence in themselves—flaunting one's expertise and accomplishments and teaching as the all-knowing, all-powerful Oz curtails learning, inhibits a student's desire to ask questions, and limits success.

Secondly, asking the student questions that bring forth the answer he or she already knows is magical. The best teacher I ever had in college, Dr. Edmund Epstein, was a brilliant scholar who could easily have made anyone (even other professors) feel like an idiot but he did just the opposite. He allowed students' brilliance to shine through by helping them to draw out the answers to their questions and gave incredibly positive reinforcement at every turn. Students left his classes energized to learn more, saying, *"I had no idea I was so brilliant and already knew the answer!"*

How do you connect with your learners?
For me, it's important to:

* Believe in students and encourage them; listen to their fears and struggles.

* Find out who they are and what they expect of themselves.

* Let them choose from various writing topics. (Students often surprise me!)

- Try to locate where the class energy is. What do they want to learn and why? What topics create lively discussions?

What attributes make you a good Trainer?
I keep focusing on the basics and try to teach them anew. I believe good teaching means constant learning and relearning.

Luisa Vercillo
Software Training Specialist

How do you connect with your learners?
When someone thanks me for having a "laid back and patient" training style it always troubles me to realize that not all adults view a training session as something positive and enjoyable. It brings to mind an experience my little sister Silvana once shared with me. She was working in a hospital for the dietary department and would visit patients to discuss their menus. As she made her rounds she wore the standard white lab coat that many health professionals wear. She was working the pediatric floor that day and entered the patient's room. The poor child cringed and hid behind his mother. The mother explained to her that her son thought she was a doctor because of her lab coat and that he associated pain with the person wearing the coat.

As a Trainer of adult learners, we do not have the advantage Silvana had that day of an insightful mother to explain her child's fear. We must instead depend on our ability to interpret body language and read between the lines of negative or indifferent comments. We need to always keep in mind that for some, a "Trainer" may bring to mind a person who presented material too quickly and did not check-in during the learning process, causing the individual to feel that he or she lacked the intelligence to grasp the material. For others, a "Trainer" may remind them of someone who turned a comment or question made in complete sincerity into a moment resulting in the learner feeling ridiculed or belittled.

Technology is increasing at an ever quickening pace. This means that training sessions are being offered more frequently. Often sessions are being led by individuals who simply "talk" the information at others as one more step in a checklist of tasks in a project roll-out. I have been fortunate because my current employer shares my perspective and fosters an environment for true learning. I still begin some sessions "talking people in off the ledge" of anxiety created by ghosts of previous Trainers, but these are becoming fewer and far between as learners share their experiences with their co-workers back in the workplace and let them know that spending time with me didn't hurt a bit.

Candace Zacher
Ph.D., Instructional Design and Research
President, The Wayfinding Group, Inc.
Instructional Designer, Performance Consultant, Organizational & Leadership Development

What attributes make you a good Trainer?
Although some people might say that it's a natural talent to be an effective Trainer, I certainly think people who have a desire to train and develop others in a business setting can do an honest assessment of their own qualities and characteristics and then design a development plan to improve their abilities in training delivery. As Aristotle said, *"We are what we repeatedly do. Excellence, therefore, is not an act but a habit."*

My response to this question is in the form of a list of key thoughts. Perhaps a few will capture your attention and you will want to see how you can apply them.

- Be prepared—Yes, you really do have to rehearse your delivery! And more than once! You should want to be the consummate Trainer for your learners. And as part of this rehearsal, you will learn the content very well and feel comfortable with it.

- Apply accelerated learning concepts when possible—If you don't know the term, you should look it up and read about it. Then ask yourself, *"How can I apply this in my delivery!"*

- Engage the learners early and often—Make certain you have activities for them to apply the content knowledge you are sharing with them. At least ask them questions periodically.

- Use plenty of examples and stories—We remember stories longer than lots of facts. And the stories can help in our application of new skills and knowledge.

- Use a compelling opener as you start your training—Remember the WIIFM principle.

- Learn the balancing act of classroom dynamics—A blend of content acquisition in multiple forms emphasizing activity-centered, flexible delivery and an atmosphere of enjoyment—this doesn't mean just having fun (training should be value-added and delivering a collection of clever gimmicks and techniques is not effective training).

Remember, training is a building block to employee accomplishment. As a Trainer, you are one link in solidifying the blocks.

How do you connect with your learners?
I know that many of the TV talk shows have a Top 10 list—well I cut that in half and here's my top 5 list!

1. Be genuine and authentic. Learners get annoyed with Trainers who they don't feel they can trust.

2. Be a good listener. One key purpose you serve is to answer questions the learner may have regarding content and application.

3. Be accessible. Share your knowledge and wisdom freely!

4. Be willing to admit you don't know something! But more importantly, tell your class that you will find out the information for them. Then please be sure to follow through on what you promised. Don't forget to do this!

5. Have a sense of humor!

BIBLIOGRAPHY

Barbazette Jean, <u>Managing the Training Function for Bottom Line Results: Tools, Models, and Best Practices</u>, Pfeiffer, 2008

Biech Elaine, <u>ASTD Handbook for Workplace Learning Professionals</u>, American Society for Training & Development, 2008

Bloom Benjamin, <u>Taxonomy of Educational Objectives Handbook 1 Cognitive Domain</u>, Longman, 1984

Bloom B. S., Hastings J.T., and Madaus G. F., <u>Handbook on Formative and Summative Learning</u>, New York: McGraw, 1969

Bolton Robert and Bolton Dorothy, <u>People Styles at Work</u>, AMACOM, 1996

Brinkerhoff Robert and Apking Anne, <u>High Impact Learning</u>, Perseus Publishing, 2001

Clothier Paul, <u>The Complete Computer Trainer</u>, McGraw Hill, 1996

Dick, W., Carey, L., & Carey, J.O, <u>The Systematic Design of Instruction. 7th Edition,</u> New York: Allyn & Bacon, 2008

Hodell Chuck, <u>ISD From the Ground Up</u>, American Society for Training & Development, 2000

Islam Kaliym and Trolley Edward, <u>Developing and Measuring Training the Six Sigma Way: A Business Approach to Training and Developing</u>, Pfeiffer 2006

Kirkpatrick Donald and Kirkpatrick James, <u>Evaluating Training Programs: The Four Levels 3rd Edition</u>, Berrett-Koehler Publishing, 2006

Kirkpatrick Donald, <u>Another Look at Evaluating Training Programs</u>, American Society for Training & Development, 1998

Knowles Malcolm, Holton III Elwood F., Swanson Richard, <u>The Adult Learner</u>, Butterwort-Heinemann Publications, 1998

Knowles Malcolm, <u>The Making of an Adult Educator</u>, San Francisco, Jossey Bass, 1989

Knox, Alan B., <u>Helping Adults Learn</u>, San Francisco: Jossey-Bass, 1986

Kolb D. A., <u>The Learning Style Inventory</u>, Boston: McBer, 1976

Kroehnert Gary, <u>100 Training Games</u>, McGraw-Hill, 1994

Lucas Robert, <u>Training Workshop Essentials: Designing, Developing, and Delivering Learning Events that Get Results</u>, Pfeiffer, 2009

Merriam Sharan and Caffarella Rosemary, <u>Learning in Adulthood</u>, Jossey-Bass Inc. Publishers, 1991

Miller H. L., <u>Teaching and Learning in Adult Education</u>, New York; Macmillan, 1964

Morris L.L., <u>How to Deal with Goals and Objectives</u>, Beverly Hills, CA, Sage, 1978

Pike Bob and Arch Ave, <u>Dealing With Difficult Participants</u>, Jossey-Bass, 1997

Pike Bob and Busse Chris, <u>101 Games for Trainers: A Collection of the the Best Activities from Creative Training Techniques</u>, Help Press 1995

Piskurich George, <u>Rapid Instructional Design</u>, John Wiley & Sons, 2006

Rae L., <u>How to Measure Training Effectiveness</u>, New York, Nichols, 1986

Renner Peter, <u>The Art of Teaching Adults</u>, Vancouver, Training Associates, 1993

Robinson Dana Gaines and Robinson James C., <u>Performance Consulting</u>, Berrett-Koehler Publishers, Inc. 1995

Robinson D. and Robinson J. S., <u>Training for Impact</u>, San Francisco: Jossey-Bass, 1989

Rothwell William and H.C. Kazanas, <u>Mastering the Instructional Design Process</u>, San Francisco: Jossey-Bass, 1992

Stolovitch Harold, <u>Telling Ain't Training</u>, ASTD Press, 2002

Worthen Blaine and Sanders James, <u>Educational Evaluation</u>, Longman, 1987

ABOUT THE AUTHOR

Sharon Johnson-Arnold has designed and developed documentation and training for Fortune 1000 companies and public sector clients for over 25 years. By functioning as the liaison between system architects and the user community, she is committed to designing training and documentation that are clear, concise, and easy to understand. Her goal is to ensure that users can clearly understand how to use software to maximize the customer's investment in technology and foster job competence and satisfaction.

Sharon Johnson-Arnold is the Principal of **TechnoWrite, Inc. (TWI).** **TWI** is a Chicago-based consulting firm specializing in technical writing, instructional design, software training, and e-learning development. TWI has been in the business of designing instructional systems to implement new software applications and Enterprise Resource Planning (ERP) systems.

TechnoWrite's team of Technical Writers, Instructional Designers, and Software Trainers are learning professionals and communicators who are trained to look at computer systems from the viewpoint of the innocent user. By translating technical concepts into user-friendly non-intimidating language, TWI participates in the entire life cycle of the training and development process—from the *inception* (analysis and design) through *implementation* (materials development and training delivery), and finally *completion* (evaluation).

Winner of the Society for Technical Communication's Achievement Award, Sharon has also served as a judge for national competitions. She holds a Bachelor of Science Degree in Business Administration (Management) from Roosevelt University and a Master of Education Degree (Instructional Technology and Adult Learning) from Loyola University. She is married to Colonel Dr. Damon Arnold, MD, MPH. They reside in Chicago.